Exploring the Ethical Dimension in Recording and Documenting Cultural Heritage

Efstratios Stylianidis

Exploring the Ethical Dimension in Recording and Documenting Cultural Heritage

Springer

Efstratios Stylianidis
Laboratory of Geoinformatics
Aristotle University of Thessaloniki
Thessaloniki, Greece

ISBN 978-3-031-80033-7 ISBN 978-3-031-80034-4 (eBook)
https://doi.org/10.1007/978-3-031-80034-4

This Springer imprint is published by the registered company Springer Nature Switzerland AG
The registered company address is: Gewerbestrasse 11, 6330 Cham, Switzerland

Excellence is not an act, but a habit.

—Aristotle

This book is dedicated to all those tirelessly working and advocating for the preservation of cultural heritage around the world, whose efforts safeguard our shared history for future generations ...

Preface

It is with great enthusiasm and a deep sense of responsibility that I present to you *Exploring the Ethical Dimension in Recording and Documenting Cultural Heritage.* This book is the culmination of extensive research, thoughtful reflection, and a profound commitment to preserving the rich tapestry of our world's cultural heritage with integrity and respect.

The journey to this book began with a simple yet profound question: How can we ensure that our efforts to record and document cultural heritage are not only accurate but also ethically sound? In an era where technology has vastly expanded our capabilities to capture and share cultural expressions, it is imperative that we navigate these advancements with a keen awareness of the ethical implications they carry.

Chapter 1: *Introduction and Scope* sets the stage for our exploration, providing an overview of the objectives and the critical need for ethical considerations in cultural heritage documentation. This chapter lays the groundwork by defining the scope and the key questions that drive our inquiry.

Chapter 2: *Recording and Documenting Cultural Heritage* delves into the methodologies and technologies used in the field. Here, we examine the tools and techniques that have revolutionized our ability to document cultural artifacts and practices, and we also scrutinize the potential ethical dilemmas that arise from their use.

In Chap. 3: *The World Cultural Heritage Context*, we expand our view to the global stage. This chapter discusses international frameworks, conventions, and the roles of various organizations dedicated to cultural heritage. It highlights the importance of a collaborative approach and the need for a unified ethical stance across borders.

Chapter 4: *Ethical Excellence* is at the heart of our discourse, where we explore what it means to achieve ethical excellence in the context of cultural heritage. Through case studies and theoretical insights, we identify principles and best practices that can guide practitioners toward making ethically sound decisions.

Chapter 5: *Ethics and Responsibility* takes a closer look at the responsibilities of individuals and institutions involved in cultural heritage documentation. This chapter emphasizes accountability, transparency, and the moral obligations that come with the privilege of preserving cultural narratives.

Chapter 6: *Trustworthy Data* addresses the crucial aspect of data integrity. In this age of information, ensuring that the data we collect, and share is reliable and trustworthy is paramount. This chapter explores strategies for maintaining data accuracy, authenticity, and security, thus safeguarding the cultural heritage we document for future generations.

Chapter 7: *Legal Aspects* explores the legal frameworks essential for the protection, documentation, and ownership of cultural heritage. It covers intellectual property rights, data protection, and informed consent, emphasizing the need for ethical practices. The chapter also addresses challenges in ownership and custodianship of artifacts, advocating for international cooperation and harmonized laws to prevent exploitation and illicit trafficking.

As you embark on this exploration, I hope this book serves as a guide and an invitation to engage in a broader conversation about the ethical dimensions of our work. Cultural heritage is more than just a record of our past; it is a living, evolving entity that shapes our present and future. Our approach to recording and documenting it must be thoughtful, respectful, and deeply ethical.

Thank you for joining me on this journey. I trust that the insights and discussions within these pages will inspire you to consider the ethical implications of your own work and to strive for excellence in preserving the cultural heritage that enriches our world.

Thessaloniki, Greece Efstratios Stylianidis

Acknowledgements

Throughout my years of work in the field of cultural heritage, I have had the opportunity to learn so much from my professional experiences. My first acknowledgment must go to those who have given me the opportunity to participate in the documentation of cultural heritage objects across various countries around the world. Each project, undertaken in diverse cultural landscapes, has deepened my understanding and appreciation of our shared global heritage. The privilege of contributing to the documentation and preservation of these priceless artifacts and sites has been both an honor and a profound responsibility.

A book like this is fundamentally shaped by practical experience in the field of cultural heritage documentation, as well as by the information—both oral and written, formal and informal—gathered from numerous and diverse sources around the world. The bibliography indirectly acknowledges these invaluable resources. In some instances, sources have been used but had to be reformulated or adapted to fit the specific content of this book. As such, they cannot be acknowledged directly. Nevertheless, the collective wisdom and expertise that have informed my work are woven throughout the text.

As the author of this book, I am deeply indebted to many colleagues and friends worldwide with whom I have become professionally acquainted. It is, however, impossible to name all those involved. I would particularly like to acknowledge the support and wealth of advice I have received over the years from my colleagues in the international community regarding the documentation of cultural heritage. The pressing task of preserving our cultural heritage has fostered the collaboration of a dedicated team—comprising archaeologists, architects, engineers, historians, and many others—working together toward this shared goal. Their passion and commitment have been a constant source of inspiration, significantly shaping the direction of my research and practice.

The content of this book has been further refined through my teaching and research experience at the Aristotle University of Thessaloniki, Faculty of Engineering, School of Spatial Planning and Development, in Thessaloniki, Greece. I am deeply grateful for the work we are doing in the Laboratory of Geoinformatics, which I have had the privilege of leading since its establishment in 2020. I thank my university for the opportunity to teach young minds and pursue my research. The dynamic academic environment, coupled with the enthusiastic engagement of my students, has consistently motivated me to explore new ideas and methodologies. Their curiosity and innovative thinking have often led to new insights and breakthroughs in my work.

I also owe a great debt to my friends in CIPA-Heritage Documentation, with whom I have shared an incredible journey spanning more than two decades. I have served CIPA in various capacities, including as a member of the Executive Committee, Secretary General, and President. I look forward to continuing to serve as Vice President during the upcoming 2024–2027 term. Since 2014, we have been organizing a summer school—3D Surveying and Modeling in Cultural Heritage—in various locations around the world. These summer schools have been not only platforms for sharing knowledge and skills but also opportunities for fostering lasting friendships and professional networks. My understanding and practice of cultural heritage documentation have been greatly enriched by the collaborative spirit and exchange of ideas that these events have fostered.

As is always the case, the author bears full responsibility for the content of this book. While there may be mistakes, I hope that they are not significant enough to detract from the experience of readers. I have made every effort to present the material as accurately and comprehensively as possible, and I welcome any constructive feedback that may help further improve the quality of this work.

I would like to express my heartfelt gratitude to my family and friends for their unwavering support and understanding throughout this long and demanding process. Their patience and encouragement have been a source of strength, allowing me to persevere and bring this book to completion. I also extend my thanks to the many unnamed individuals who, in one way or another, have contributed to this project. Every interaction, no matter how small, has played a role in shaping the final outcome of this work.

Thank you all for being part of this journey.

Contents

Abbreviations

AIC	American Institute for Conservation
APT	Association for Preservation Technology
FIM	Forward Intersection Method
GDPR	General Data Protection Regulation
GHF	Global Heritage Fund
GIS	Geographic Information Systems
GNSS	Global Navigation Satellite System
IBM	Image-Based Modeling
ICA	International Council on Archives
ICOMOS	International COuncil on MOnuments and Sites
IPRs	Intellectual Property Rights
ISPRS	International Society for Photogrammetry and Remote Sensing
LiDAR	Light Detection And Ranging
MPC	Method of Polar Coordinates
MVS	Multi-View Stereo
RBM	Range-Based Modeling
RICS	Royal Institution of Chartered Surveyors
SfM	Structure from Motion
SHM	Structural Health Monitoring
SMEs	Small and Medium-Sized Enterprises
TLS	Terrestrial Laser Scanning
ToF	Time of Flight
UN	United Nations
WHC	World Heritage Centre
WHF	World Heritage Fund
WHS	World Heritage Sites
WMF	World Monuments Fund
WOS	Web of Science

Chapter 1
Introduction and Scope

Abstract Cultural heritage represents humanity's achievements and serves as a bridge between the past and future, offering knowledge and tools for enlightenment. It plays a vital role in shaping community identity and history, despite being vulnerable to natural disasters and human actions. Protecting cultural heritage, from small artifacts to large monuments, is crucial due to their economic, social, and spiritual significance. Abandonment and destruction, driven by economic, environmental, and social factors, pose significant threats. Preservation requires interdisciplinary efforts and documentation, involving local communities, governments, and international organizations like UNESCO. Digital technologies and collaborative approaches enhance the preservation and appreciation of cultural heritage globally.

1.1 Introduction

Cultural heritage is an irreplaceable property of humanity. It depicts the achievements of humankind in every level of human life over centuries across the world and it serves a guiding path from centuries past into the future. Cultural heritage provides the knowledge, and also essential tools for the enlightenment of the generations to come.

The relationship between people and cultural heritage is profoundly significant. The interconnection influences a community's identity, rights, history, and future. From small artifacts to huge monuments and sites, they all bear witness to these evolving relationships between people and their cultural, environmental and social ecosystem. Unfortunately, the fragility of cultural heritage is frequently threatened by either natural disasters or human-made destruction.

For many, ancient objects, monuments and sites make substantial contributions to human knowledge. In the realm of science, artifacts and findings illustrate how people, communities, and even the Earth itself have transformed over time, offering insights into how our ancestors adapted. This knowledge can be instrumental in guiding climate change management and fostering sustainable practices that resonate with our historical context.

E. Stylianidis, *Exploring the Ethical Dimension in Recording and Documenting Cultural Heritage*, https://doi.org/10.1007/978-3-031-80034-4_1

Palaeontological, archeological even architectural artifacts and structures are irreplaceable, guardians of our shared heritage. This is a reason making their protection crucial. Monuments and sites extend beyond their physical presence. They hold economic, social, and spiritual significance, enriching communities' understanding of their identity. They connect people to a specific place and time, creating a fertile environment and opportunities for gaining new knowledge and understanding one another and ourselves.

Moreover, the synergy between innovating and understanding cultural heritage can offer remarkable prospects in the contemporary context. It provides a dynamic platform where the past intersects with the future, fostering creativity and progress, and is a vehicle for employment opportunities.

Respecting the past brings immeasurable benefits to individuals and communities. A thorough understanding of the past is not only valuable. It is essential to building a sustainable future for humanity.

1.2 Abandonment of Cultural Heritage

Abandonment refers to the act of leaving something behind. In the context of cultural heritage, it can be a complex and often contradictory term. It is important to consider how cultural heritage, esteemed for its significant and recognized value, can be the subject of abandonment.

Despite this, abandonment can occur in various locations. In many traditional villages and historic cities around the world, historic sites are abandoned, and historic buildings and monuments suffer neglect, leading to deterioration. There are many reasons why this valuable cultural heritage is being neglected. Economic, environmental, and social factors all contribute to the abandonment of our cultural property.

Abandonment is a preservation concern, particularly for valuable objects of cultural significance, artistic works, historic buildings, monuments, and sites that require ongoing, detailed, and specific repair. Repairing such buildings can be costly, often exceeding the value of the building. It is important to consider the cost of repair in relation to the value of the building.

In many cases, the owner will have the choice of either tearing down the building or leaving the building vacant. In some other cases, owners who are faced with the high cost of repairing a building may find that they have to make a difficult choice. They may have to make difficult decisions such as whether to demolish the building or leave it vacant. This dilemma highlights the balance between the preservation of cultural heritage and the economic feasibility of restoration efforts.

Addressing the issue of cultural heritage abandonment requires a multifaceted approach that considers not only the architectural, historical, and cultural significance but also the economic and practical challenges associated with preservation. To find

Fig. 1.1 Abandoned house in a Greek city. (*Source* Pixabay)

sustainable solutions and ensure that valuable cultural heritage is not abandoned, collaborative efforts involving local communities, governments, and preservation organizations are essential.

Abandoned buildings (Fig. 1.1) and sites can evoke negative emotions and melancholy in urban dwellers, especially those with significant cultural value. They can also be perceived as dangerous. They can attract undesirable or even illegal activities. However, they also have a great potential to serve as creative spaces for neighborhoods and local communities. Historic structures in these places can provide a link to the past. They can also provide a unique setting for various activities.

Abandoned cultural heritage sites and monuments can deteriorate over time due to neglect or vandalism, making them vulnerable to plunder and illegal trafficking. This can result in the loss of valuable historical and cultural information and artifacts. It can also lead to the degradation of local cultural traditions and identities.

Several international organizations and initiatives have been established to address the problem of abandoned cultural heritage. These include the UNESCO World Heritage Centre (WHC) and the Global Heritage Fund (GHF) of the World Monuments Fund (WMF). The WHF is UNESCO's dedicated fund for the protection of the world's cultural and natural heritage, especially those of outstanding universal value. It was established in 1977 under Article 15 of the World Heritage Convention. The WHF plays a vital role in supporting efforts to conserve and protect significant global heritage sites. Based on the principle that the protection of cultural heritage can act as a catalyst for responsible social and economic development, the Global Heritage Fund (GHF) was established. The GHF, which is affiliated with the WMF, aims to use heritage conservation to promote sustainable development practices. It promotes a balanced approach to development that respects the past while building for the future by demonstrating how the preservation of historic and cultural treasures can contribute to the well-being of communities.

1.3 Destruction of Cultural Heritage

On the other hand, cultural heritage can be destroyed not only by abandonment but also by natural phenomena and human-made causes. Many cultural heritage sites around the world remain under threat. These include World Heritage Sites (WHS), which are at risk of damage from climate change and rising sea levels. Cultural heritage destruction refers to the intentional or unintentional damage or loss of valuable and significant cultural artifacts, sites, monuments, buildings, and more. It can occur for a variety of reasons, including natural disasters, conflict, war, terrorism, vandalism, urban development, or neglect.

The destruction of cultural heritage can have significant social, economic, and cultural impacts. These impacts can include the loss of valuable historical and cultural data, the erosion of local cultural identities and traditions, and the loss of benefits from tourism and other economic activities.

The destruction of cultural heritage is a pressing global issue that demands the attention and cooperation of governments, international organizations, stakeholders, and communities worldwide. Several international organizations and initiatives, such as the WHC of UNESCO, ICOMOS, and Blue Shield International, have been established to address this problem.

Tourism is often viewed as a significant source of income and employment for countries due to its labor-intensive nature. Cultural and historical legacies are also major attractions for tourists in many destinations, making them a crucial resource for generating current and future tourism. However, conflicts around the world are endangering cultural heritage, which in turn damages the future development potential of affected economies.

Terrorism, conflict, and political violence make international tourism highly sensitive. As a result of tourists' desire for safe destinations, if people perceive any risk, they will avoid that particular destination. (Neumayer 2004) has examined the impact of terrorism and political instability on international tourism. (Neumayer and Plumper 2016) have analyzed spatial spillover effects in international tourism as a consequence of transnational terrorist attacks, and specifically, they hypothesize that attacks executed in Islamic countries on citizens from Western countries will generate different kinds of spatial spillovers. However, the negative effects of these extreme socio-political conditions go beyond the short-term decline in inbound tourism.

1.3.1 Natural Disasters

Natural disasters pose a significant threat to the preservation of cultural heritage. Earthquakes, floods, hurricanes, and wildfires are just a few examples of disasters that occur around the world. Factors such as geographic location, geological formations, and other environmental elements contribute to the likelihood of a natural disaster occurring. At times, these events can occur suddenly or with very little warning and can cause irreparable damage and destruction to monuments and archaeological sites.

The impact of a natural disaster on a cultural heritage site can be devastating. It can cause severe structural damage to historic buildings, monuments, and sites. Cultural heritage sites in coastal areas are particularly vulnerable to erosion and submersion as a result of sea level rise and storms. Heavy rainfall and flooding can cause water damage to the interior of buildings, monuments, and archaeological sites. Particularly vulnerable to moisture are frescoes, mosaics, wooden artifacts, manuscripts, etc.

In 2017, for the first time, a group of experts (Pavlova et al. 2017) analyzed the estimated exposure of 981 World Heritage properties worldwide to geological hazards and the risk awareness of their managers. The study focused on the physical exposure of heritage sites to four main geological hazards: tsunamis, landslides, earthquakes, and volcanic eruptions. The results of this study showed that a range of 39% according to open access data, to 46% according to site managers, of World Heritage properties are exposed to at least one of these four geological hazards. The most common natural hazards affecting World Heritage properties are earthquakes and landslides. Volcanic eruptions and tsunamis are less common. The most vulnerable regions, where more than half of the sites are exposed to at least one geohazard, are Asia and the Pacific, and Latin America and the Caribbean.

Wildfires

Climate change, coupled with periods of drought, can give rise to wildfires, leading to the destruction of priceless artifacts, monuments, and sites. Wildfires can have various origins, including arson, drought, lightning, human error, or even short circuits. In such sudden events, wildfires can result in significant losses of artifacts, archives, collections, historic buildings, and even entire historic cities. The intense heat and flames can cause wooden structures to burn, stone structures to crack or shatter due to thermal stress, and metal elements to warp.

Notre-Dame in Paris, France, is a medieval Catholic cathedral consecrated to the Virgin Mary, renowned for its status as an eminent example of French Gothic architecture. Designated as a UNESCO World Heritage Site in 1991 as part of Paris, Banks of the Seine (UNESCO 1991), Notre-Dame faced a devastating fire on April 15, 2019. The monument's spire collapsed (Fig. 1.2), and a significant portion of its roof was destroyed, with upper walls suffering severe damage. While some works

Fig. 1.2 Notre Dame's spire image during the 15th April 2019 fire. (*Source* Wikimedia | CC BY-SA 4.0)

of art and religious relics suffered from smoke damage, many were successfully transferred to a safe location. The French government has committed to completing the reconstruction of the cathedral within five years.

In their study, (Jacquot and Saleri 2023), presented the issues, methods and experiments to enable the acquisition of data and the creation of digital models of the wooden structures, also known as the forest, of Notre-Dame de Paris in different temporal states. This experiment proposes the necessary qualities to enrich the future ecosystem of analysis and multidimensional storage for the restoration and conservation of Notre-Dame.

According to (Luca 2023), the Notre Dame fire highlighted the critical role of digital data in cultural heritage research and the need for comprehensive, multidisciplinary data interpretation. As a result, an initiative has been launched to create a digital ecosystem that represents both the tangible object and the evolving knowledge surrounding it. This initiative integrates digital humanities, heritage science, and computer science, emphasizing an ecosystem tailored for human-led data collection and multi-dimensional data integration.

Wildfires can cause significant damage to natural landscapes, which in turn affects cultural heritage sites located in or near the affected regions. Changes in vegetation, soil composition, and erosion can all contribute to the destruction of monuments and sites and the degradation of a region's historical context. A devastating example of these impacts was seen during the 2019–2020 Black Summer fires in eastern Australia, which had a profound impact on numerous Indigenous heritage sites. This included the destruction of invaluable elements such as ancient rock art and landscapes. The preservation of Indigenous cultural heritage is an integral part of the multifaceted fabric of Indigenous cultures, and plays a central role in fostering and maintaining a deep connection to the land.

Earthquakes

Earthquakes occur as a result of the collision of tectonic plates, and thus the main consequence of an earthquake is that the energy released in the form of waves causes shaking. Earthquakes threaten thousands of cultural heritage sites around the world, monuments and sites, and valuable objects in museums, to name a few. Earthquakes unsettle the stability of historic buildings and can cause severe cracks and damage in various parts of the buildings. This natural disruption can lead to the destruction of art, artifacts, and many other cultural heritage objects. It can also, under certain circumstances, lead to the looting of cultural heritage objects.

Bagan built heritage is located in central Myanmar, in Southeast Asia, and it was developed between the 9th and 13th centuries as the capital of the Myanmar kingdom. It consists of both secular and religious architecture, and the site has a complex variety of architectural types, including monasteries, underground structures, royal palaces, local housing, stupas, and temples (Mezzino et al. 2017). The site is included in the UNESCO World Heritage List (UNESCO 2019). In recent years, this exceptional

World Heritage Site has been exposed to an increasing number of natural hazards, including earthquakes, which have affected its built structures. According to the Bagan Disaster Risk Management Plan (MoCRA 2017), on August 24, 2016, an earthquake of magnitude 6.8 struck near Chauk, Myanmar, at a depth of 82.0 km. The earthquake damaged more than 350 monuments in the Bagan Cultural Heritage Site. The Sulamani temple was one of the most famous damaged during this earthquake (Fig. 1.3).

After the earthquake of 2016 in Myanmar, an interdisciplinary group of experts (Costamagna 2020) performed a study using advanced non-destructive techniques for the diagnosis of historic buildings, targeting the Loka-Hteik-Pan temple in Bagan. The monument was significantly damaged by the earthquake, losing the upper part of the tower, as happen with many other temples in this specific area. The authors presented the preliminary results of the tests in order to provide a knowledge base, helpful to support the interventions for preserving the monument and its heritage.

The powerful shaking of an earthquake can severely damage or even collapse monuments and historic structures. Earthquakes can damage artifacts and archives stored in museums or libraries. Fragile objects such as ceramics and artwork can be broken by the impact of seismic waves during a strong quake. Seismic activity can trigger landslides and rockfalls, which can cause burial or severe damage to cultural heritage sites.

The 2016 earthquake in Central Italy caused severe economic losses and a large number of casualties (303 deaths). Several municipalities in central Italy, including Amatrice, Accumoli, Arquata, Norcia and Visso, registered severe damage to

Fig. 1.3 The Sulamani temple was one of the most famous to be damaged. (*Source* Wikipedia | CC BY-SA 4.0)

the building stock, including the 13th century Basilica of Saint Benedict in Norcia (Ferracuti 2022). A year earlier, the 2015 Gorkha earthquake was the most powerful and devastating to hit Nepal since 1934. (Weise et al. 2018) recorded that the earthquake damaged 663 monuments and caused the collapse of 190 other vernacular and historical monuments.

Tsunamis

Tsunamis are massive waves caused by earthquakes or volcanic eruptions that occur under the sea. In the depths of the ocean, these waves are not high enough to cause much of a problem on the surface of the earth. However, as the waves move along the land, they become much higher as the depth of the sea decreases. When the waves reach the shore, they can be damaging and dangerous to human life, infrastructure and cultural heritage where it exists.

On March 11, 2011, an earthquake and tsunami struck the northeast coast of Japan. The catastrophic tsunami that followed claimed many thousands of lives and left millions without power and water. These natural disasters left behind severely damaged monuments and cultural landmarks. According to the Japan Agency for Cultural Affairs, more than 700 national landmarks were damaged, most of them in the Tōhoku and Kantō regions. A large number of other traditional buildings and historic urban landscapes were damaged in places such as the coastal city of Kesennuma in the Tōhoku region. The earthquake and tsunami were so powerful that countless historic buildings became damaged, deformed, and suffered partial or complete collapse (WMF 2012).

(Okamura 2015) described that on 11 March 2011, Japan experienced one of the largest seismic events the world has seen in generations. It affected a large area covering parts of the Tohoku and Kanto regions and stretching more than 500 km from north to south. The damage was particularly severe in the prefectures of Miyagi, Iwate, and Fukushima. The excavation site of the medieval fortified village of Niidatate, the city of Minamisanriku in Miyagi Prefecture, was destroyed. The entire village at the foot of the hill was swept away by the tsunami (Fig. 1.4).

Local knowledge and culture are important in how communities respond to and recover from disasters. This importance is recognized in the Sendai Framework for Disaster Risk Reduction 2015–2030, developed by the United Nations to prevent new and reduce existing disaster risks (UNODRR 2015). For the Great East Japan Earthquake and Tsunami of 2011, (Iizuka 2021) conducted a study to understand the role of local performing arts in post-disaster recovery. To gain insights into the experiences and perspectives of local people, the author introduced a qualitative descriptive methodology. The findings of this study were that local performing arts are part of people's lives and are based on knowledge and experiences that have been passed down from generation to generation. Although people experienced the tsunami, they gradually began to perform again within the community as a means of regaining their identity and encouraging themselves and others.

Fig. 1.4 The entire village of Niidatate, Japan, at the foot of the hill was swept away by the tsunami (2011). (*Source* (Okamura 2015)

Floods

Flood-related impacts constitute one of the most frequent, serious and damaging threats of cultural heritage. This type of impacts are enormous and the scenes where monuments or works of art are flooded over several days are scary. The objects are covered in water, mud, or even sewage waters. Certain cultural heritage objects may not only be completely destroyed by a flood, but records reporting their existence may be destroyed or severely damaged too (Deschaux 2017).

Flooding poses a serious threat to cultural heritage. It damages the structural condition of historic buildings, monuments, and sites. The water causes erosion of foundations. This can lead to collapse and accelerate the deterioration of materials. Flooding also erodes and deposits soil, burying or washing away archaeological sites and artifacts from archaeological excavations. Obviously, this can lead to the loss of valuable historical information, as well as the damage or destruction of archives and documents.

Google Arts and Culture sheds light on 10 cultural sites that have been bombed, flooded, destroyed and built over. The case of flooded Florence in 1966 is listed first: "*At dawn on November 4th, 1966, a flash flood that swelled the river Arno hit the Renaissance capital at 37mph. Historic libraries, churches and art galleries were inundated with water rising up to 22ft (6.7m high), mixed with heating oil, mud and sewage. Countless thousands of priceless books, maps and works of art were damaged or destroyed, with images flaking off painted panels and frescoes. Fortunately, thousands more were rescued by volunteer 'Mud Angels'. Among the*

Fig. 1.5 Manuscripts from the National Library being washed and dried in the boiler room of Firenze Santa Maria Novella railway station, after the flood. (*Source* UNESCO/Dominique Roger | CC BY-SA 3.0)

famous art casualties was the 13th century Cimabue crucifix from the Santa Croce basilica. Conservation continues today" (Google 2020).

Figure 1.5 illustrates people washing and drying manuscripts from the National Library in the boiler room of Firenze Santa Maria Novella railway station after the devastating flood of 1966 in Florence, Italy.

Italy has many historical monuments and art works. (Lanza 2003) has studied the threat to the cultural heritage of the historic center of the city of Genoa as a result of flooding. The area, due to the lack of defensive and/or preventive structures that are now hardly compatible with the ancient urban structure, is exposed to the risk of failure of the urban drainage system. In order to derive a map of historically flooded areas, an extensive study of the available records of flood events over the last 100 years was carried out.

1.3.2 Human Threats

Cultural heritage monuments and sites around the world are increasingly threatened by natural disasters. However, negative, harmful, and violent human interventions also pose a threat to it. There are numerous examples from various regions of the world that illustrate the devastation caused by ongoing conflicts or acts of terrorism. Human-made threats such as war and terrorism pose a significant threat to our cultural heritage due to their destructive impact on humanity.

It is crucial to preserve our shared cultural heritage. In regions of conflict, monuments, historic sites, and artifacts are often deliberately destroyed, erasing tangible links to our past. The nature of warfare can lead to the loss of irreplaceable cultural treasures, as cultural property is often caught in the crossfire.

Additionally, damage to cultural heritage can result from neglect of safety regulations and human recklessness. Tragic outcomes occur when individuals fail to recognize the importance of preserving cultural heritage, leading to irreparable damage. In some cases, the unintentional loss or damage of cultural property has resulted from pursuing economic interests or disregarding established conservation guidelines.

To address these challenges, societies must prioritize the protection of cultural heritage. This involves not only documenting and preserving cultural artifacts but also promoting a global understanding of the significance of cultural diversity and historical continuity. International cooperation and diplomatic efforts are crucial to mitigate the impact of armed conflict and terrorism on cultural heritage and to ensure the preservation of these valuable aspects of human history for future generations.

War and Terrorism

Over the past decades, the world has witnessed an increasing number of terrorist attacks on cultural heritage that have destroyed it. These attacks have occurred in countries affected by armed conflict. They have also been accompanied by organized looting, trafficking, and sale of cultural property. As part of a carefully planned strategy to intimidate governments and populations, terrorist groups use these illegal actions. The objective of terrorists is the generation of income for their groups worldwide. This income is then used to increase recruitment and to support illegal operations.

Terrorism has its roots in the historical, ideological, and political rhetoric surrounding cultural heritage. In her book, "*Destruction of Cultural Property as a Weapon of War: ISIS in Syria and Iraq*", (Turku 2018) aims to improve the understanding of cultural heritage in armed conflict and its relevance to counterterrorism and peacebuilding strategies. It also makes the case that the systematic theft and destruction of cultural heritage by ISIS has become a tactic of war. The book provides a historical, political, and legal analysis of the behavior of radical groups toward cultural heritage objects, and it interprets this behavior as part of their terror posture and military operations.

Afghanistan, Iraq, Libya and Syria (Fig. 1.6) have experienced war and terrorism. Cultural heritage has been destroyed and human dignity has been insulted by these heinous practices. Unfortunately, many other countries could be added to this list, where acts of purposive destruction not only harm cultural assets but also target free thinkers and affect people on social and humanitarian levels.

Fig. 1.6 Destruction of historic buildings in Syria. (*Source* Pixabay)

During the Russian invasion in Ukraine and the subsequent war, hundreds of cultural heritage sites were damaged. Since February 24, 2022, UNESCO has verified damage to 149 religious sites, 249 buildings of historical and/or artistic interest, 33 museums, 33 monuments, 18 libraries, 1 archive and 2 archaeological sites (UNESCO 2025). The affected sites are located in various regions of Ukraine, including Chernihiv, Kyiv, Kharkiv, Zaporizhzhya, Zhytomyr, Donetsk, Luhansk, Sumy, Mykolaiv, Vinnytsya, Odesa, Dnipropetrovs'k, Kherson, L'vin and Poltana. The majority of the damage is in the Donetsk Region.

To address the threat to cultural heritage during armed conflict, the international community has taken several measures. A framework for protecting cultural heritage during war was established by the 1954 Hague Convention for the Protection of Cultural Property in the Event of Armed Conflict (UNESCO 1954). The Convention aims to protect cultural property, which is explicitly defined in Article 1 ... *The term 'cultural property' shall cover, irrespective of origin or ownership:*

(a) *movable or immovable property of great importance to the cultural heritage of every people, such as monuments of architecture, art or history, whether religious or secular; archaeological sites; groups of buildings which, as a whole, are of historical or artistic interest; works of art; manuscripts, books and other objects of artistic, historical or archaeological interest; as well as scientific collections and important collections of books or archives or of reproductions of the property defined above;*

(b) *buildings whose main and effective purpose is to preserve or exhibit the movable cultural property defined in sub-paragraph (a) such as museums, large libraries and depositories of archives, and refuges intended to shelter, in the event of armed conflict, the movable cultural property defined in sub-paragraph (a);*
(c) *centers containing a large amount of cultural property as defined in sub-paragraphs (a) and (b), to be known as 'centers containing monuments'.*

Others

However, there are cases in which the human factor, whether by negligence or by omission, can have a negative impact on monuments, sites, or other objects of cultural heritage. Problems in connection with the lack of safety measures or the lack of care in certain procedures can be the cause of fires or explosions. It is of vital importance to pay attention to and take care of the protection of our cultural property, whether it is located in museums, historical buildings, archaeological sites, or any other place.

On August 4, 2020, in Beirut, Lebanon, a devastating explosion killed at least 200 people and injured about 5,000 others. But what caused the explosion? Lebanese Prime Minister Hassan Diab blamed the blast on 2,750 tons of ammonium nitrate. The Lebanese government resigned. This huge amount of ammonium nitrate had been in unsafe storage in a warehouse at the port. In principle, ammonium nitrate is relatively safe if it is stored properly, but large quantities of the material in storage over a long period can lead to decomposition. A large fire in the port of Beirut preceded the disaster. There was a large explosion at the beginning, followed by a series of smaller explosions. Shortly after that, there was a colossal explosion that sent a mushroom cloud into the air and a supersonic blast wave through the city of Beirut. The blast leveled buildings near the port. It also caused extensive damage to much of the rest of Beirut. The explosion irreversibly damaged much of the historic city (Fig. 1.7), including Gemmayzeh and Mar Mikhael neighborhoods.

By applying digital technology in post-disaster areas, (Kallas and Napolitano 2023) demonstrated an efficient workflow for accelerating the assessment process of damaged historic buildings from the 2020 Beirut blast. This was the first time that Beirut's historic buildings had been the subject of large-scale digitization. Emphasizing the timely benefits of disaster recovery and improved decision-making in a similar emergency, the authors described the implementation of 3D image-based modeling. The 3D documentation provided valuable data and information to heritage experts and engineers by creating accurate models of the damaged historic buildings in Beirut.

Fig. 1.7 Destruction of historic buildings in Beirut, Lebanon, after a devastating explosion on August 2020. (*Courtesy* Joe J. Kallas)

1.4 Cultural Heritage Preservation

1.4.1 Conservation, Preservation or Restoration?

The terms *conservation*, *preservation*, and *restoration* are often used interchangeably with each other. It may be that there is no problem with this in day-to-day use. However, these three terms have different meanings depending on the context and purpose, and in the following, we will explain each of the terms and provide examples to illustrate the differences between them.

Conservation is an interventive process that involves actions aimed at preserving cultural heritage in the long term. These actions include observation, examination, recording, documentation, treatment, and preventive care. Conservation is the practice of preventing further deterioration of an object. It is performed by a professional known as a conservator. The conservator may use specialized techniques to clean the object or to remove agents that are causing damage to it. Conservation does not aim to restore the object to its original state, but rather to treat it. It has a clear objective: to treat the object so that it can be safely managed, exhibited, or stored without risk of further damage. For example, to ensure safe handling and storage, a conservator may replace a damaged binding on a historic atlas of maps. The binding of the atlas will have been preserved, and a record of the entire process will have been kept. In addition, conservators adhere to a strict set of ethical principles, including the principle of *first, do no harm*, as in the case of restoring World Heritage monuments and sites (Fig. 1.8).

Preservation is the practice of maintaining objects through the provision of a constant storage or display environment to minimize further damage or deterioration. The term preventive conservation is often used to be more precise. Deterioration can be caused by environmental factors, inherent vice, or the instability of the synthesis of the object. For example, historical maps are typically made of paper, which can lead to an inevitable deterioration over time. Preservation, or preventive conservation, is

Fig. 1.8 Restoration work in progress at World heritage ancient Parthenon, Athens, Greece. (*Source* Photo by Walter Martin Unsplash)

usually accomplished through the control of humidity, light, temperature, and other potentially damaging factors to slow the process of decay.

Restoration refers to a permanent change in a cultural heritage object, making it crucial in interpreting and retelling history. Cleaning and replacing a significant number of parts can alter the historic integrity of the object, and once the object is altered, its origin is also changed. Restoration should only be undertaken after careful consideration of ethics, the probability of success, and the recommended treatment plan. Restoration must be carried out by professionals who understand the object's design and use safe practices, cleaners, or chemicals; otherwise, irreversible damage may occur. It is important to note that after restoration, the object's original condition is no longer intact, and its function and research value may be diminished. This requires careful consideration and thorough documentation of every step in the restoration process.

In their study, (Li and Tang 2024) aim to identify contemporary principles for conserving cultural heritage by reviewing its history, starting from the 18th century, and examining practices, international documents, and literature on the topic. The aim is to prevent damage to cultural heritage through improper conservation practices. The study revealed that cultural heritage conservation involves both tangible and intangible aspects, including particularly important objects and those closely related to daily life. Additionally, document-based restoration is necessary for conservation, and it is important to identify their value in modern times. Third, it is essential to integrate cultural heritage conservation into the context of sustainable development goals to successfully balance the relationship between the built and natural environment. Fourth, the mobilization of the public to participate in cultural heritage conservation enables the implementation of conservation to meet public expectations and can positively influence public attitudes. Fifth, cultural heritage conservation is a global responsibility that requires cooperation. Sixth, interdisciplinary methods for cultural heritage conservation have been developed and could be applied. However, their use should be limited to preserve authenticity and integrity.

1.4.2 *The Importance of Cultural Heritage Preservation*

The importance of preserving cultural heritage is immense. It serves as a bridge that connects people and communities around the world. The most effective approach to preserving cultural heritage is to share it widely. This allows the message of its importance to resonate with people and communities around the world. This sharing fosters an appreciation for the emotions, values, and unique historical narratives that are embedded in the cultural heritage of groups, communities, and nations.

The cultural heritage and the history of a collective entity have a deep emotional and intrinsic value and constitute a distinctive identity that serves as a recognizable feature when it is introduced to the world. This identity forms the backbone of efforts to preserve cultural heritage. It encompasses both tangible and intangible elements that have cultural significance and historical value. Cultures and their heritage not only reflect, but also influence, beliefs and values, and play a central role in the formation of the national identity of a people. By ensuring the continuity of its rich cultural tapestry, cultural heritage preservation becomes a guardian of the integrity of the people.

Intangible cultural heritage is equally important. Its value lies not only in the cultural manifestations themselves, but also in the wealth of knowledge and skills that are passed down from one generation to the next. This transmission of intangible elements becomes a conduit for the maintenance of traditions, the preservation of expertise, and the fostering of a sense of shared identity across time.

The need to preserve cultural heritage is obvious. Professionals and experts are dedicated to using every available method to achieve this goal. Throughout the world, the critical importance of recording and documenting cultural heritage is underscored by numerous documents endorsed by international organizations. A prime example is the set of principles established by ICOMOS. These principles emphasize the need for a systematic and comprehensive approach to cultural heritage preservation.

The preservation of cultural heritage has been transformed by the introduction of new technologies and tools. These innovations, ranging from tablets and computers to sensors and smart software tools, have revolutionized the conservation paradigm, providing new perspectives and opportunities. Technological advancements are utilized not only for documenting cultural heritage but also for increasing public engagement. This ensures that preservation efforts are relevant to diverse audiences in a rapidly changing world.

The preservation of cultural heritage is important for a variety of reasons, both tangible and intangible (Fig. 1.9). There are several key reasons why cultural heritage preservation is important to individuals and humanity.

Cultural heritage plays an important role in shaping and preserving the unique identity of communities and nations, distinguishing them, and facilitating global recognition. It serves as a tangible embodiment of history, traditions, and values. It fosters a deep sense of belonging and pride. Providing a vital link to the past and a sense of continuity and tradition, cultural heritage holds immense emotional and historical value for individuals and communities. Preservation efforts enable people to understand and appreciate their roots while deepening their knowledge of their own history through the preservation of monuments, sites, artifacts, and traditions. In addition, the preservation of cultural heritage contributes to the maintenance of the integrity of communities and societies. It prevents the loss of cultural diversity and promotes a more interconnected global community that celebrates diverse cultural expressions by protecting distinctive practices, languages, and customs. The transmission of intangible cultural heritage, such as knowledge, skills, and traditions, from one generation to the next ensures the continuity of valuable practices, fostering a sense of identity while adapting to changing circumstances. Cultural heritage fosters mutual respect and common ground among diverse cultures and nations, acting as a bridge for international cooperation and understanding. Collaborative efforts to preserve cultural heritage encourage the exchange of ideas and promote a global appreciation of the richness and diversity of human heritage. International organizations such as ICOMOS and committees such as CIPA-Heritage Documentation

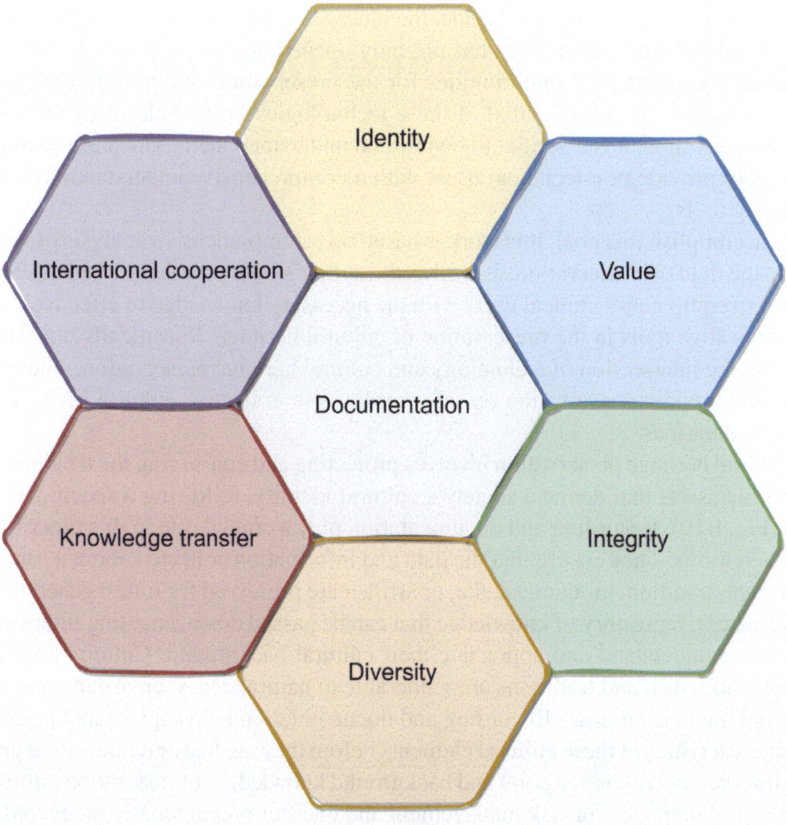

Fig. 1.9 Cultural heritage preservation is important for several reasons

emphasize the importance of documentation and record keeping and stress the need to make information about cultural heritage accessible for research, and education, and to benefit future generations.

1.4.3 Preservation Requires Recording and Documentation

Cultural heritage is an invaluable testimony to humanity's achievements over the centuries and serves as a reflection of society's accomplishments. Recognizing the importance of preserving this heritage, stakeholders are actively exploring various methods for its protection. The importance of documenting cultural heritage has been emphasized in numerous published literature and documents, such as the Burra Charter. However, the evolution of human development and the advent of new tools and technologies have had a significant impact on the concept of heritage conservation.

In this rapidly changing landscape, the emergence of technologies such as computers, sensors, and digital tools has not only opened new avenues but also created presented unprecedented opportunities for the preservation of cultural heritage. In order to realize the full potential of these technologies in the field of cultural heritage conservation, it is essential to review and understand them. The purpose of this work is to provide non-technical users with a comprehensive understanding of the emerging tools.

To accomplish this goal, this work is based on a comprehensive analysis of literature in the field of conservation, as well as the author's own experiences and research. It aims to equip non-technical users with the necessary knowledge to effectively utilize innovative tools in the preservation of cultural heritage. Specifically, this study explores the intersection of technology and cultural heritage preservation, contributing to the broader conversation on safeguarding our collective cultural heritage for future generations.

Cultural heritage preservation involves protecting and conserving the tangible and intangible assets that define a society's cultural identity, following a specific workflow (Fig. 1.10). Recording and documentation play a crucial role in this process for several reasons. They ensure that the data and information collected about a particular culture, tradition, monument, site, or artifact are preserved for future generations. This creates a repository of knowledge that can be passed down, enabling future generations to understand and appreciate their cultural background. Cultural artifacts, monuments, sites, and traditions are vulnerable to natural decay, environmental factors, and human activities. Recording and documentation help capture and preserve the characteristics of these cultural elements before they are lost or damaged. In practice, it serves as a reference point and background knowledge for restoration efforts in the event of damage. For risk management and disaster preparedness, the recording and documentation of cultural heritage are essential for its preservation. In the event of natural disasters, conflicts, wars, or other emergencies, having detailed records is

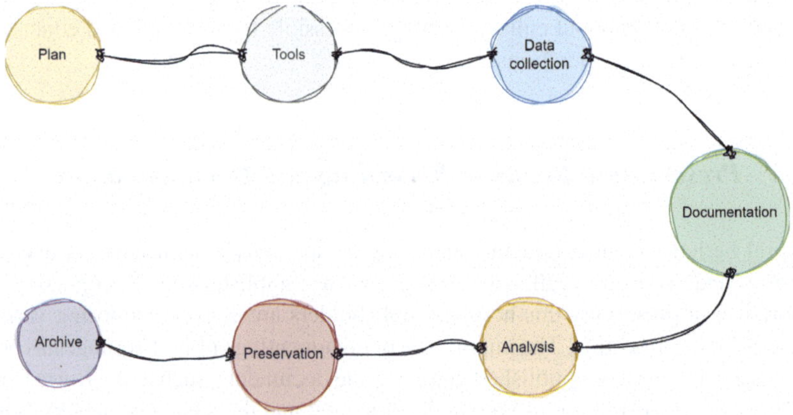

Fig. 1.10 Workflow for cultural heritage preservation

a valuable asset. It supports damage assessment and planning for the recovery and restoration of cultural heritage. Today, digital technologies are playing a catalytic role in the recording, documentation, and preservation of cultural heritage. Digital recording and documentation allow for widespread dissemination, awareness, easy access, and the creation of virtual models and archives, ensuring that both data and information are not only safely stored, but also readily available to a global audience.

1.4.4 An Essential Coalition

A coalition is an alliance or partnership formed by individuals, groups, organizations, or even nations, with a common purpose of working together to achieve shared goals or address common issues. Building a coalition involves uniting diverse entities, which may have different backgrounds, resources, expertise, or perspectives. The significance of building coalitions lies in the recognition that collective action often yields a greater impact and influence than individual efforts.

Preservation is an interdisciplinary field that requires consideration of various disciplines, including arts, humanities, science, and technology, depending on the specific application sector (Fig. 1.11). Preservation intervention should be performed by preservation professionals and experts, and the extent and specifics of the intervention should not be decided by a single individual. In practice, preservation is not solely a technical activity but rather a cultural heritage intervention that utilizes technical means.

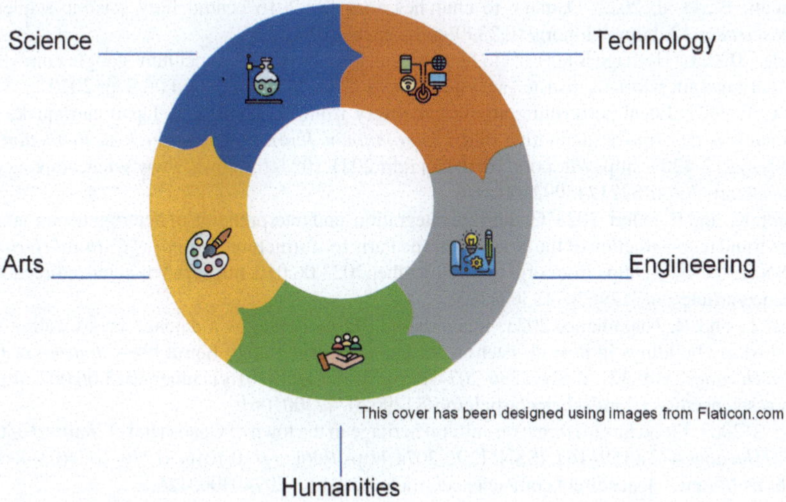

Fig. 1.11 The coalition in cultural heritage preservation

Therefore, it is crucial to plan preservation interventions on objects, regardless of their size, optimally. This involves providing interdisciplinary teams with as much information as possible about the object. The team requires time to organize the various activities, taking into account that they originate from different fields. It is important to consider the history of the object, its historical context, the available technology and materials at the time of creation, the subsequent use, and the overall conditions under which it remained.

For this reason, the contributions of many professionals and experts are crucial to conservation. Recording and documenting an object, as well as studying its structure and materials, require the active participation of all relevant scientific disciplines. The goal is to identify any changes, deterioration, or decay the object may have undergone. Based on the scientific information obtained, the appropriate conservation method can be determined.

References

Costamagna, E., et al. 2020. Advanced non-destructive techniques for the diagnosis of historic buildings: The Loka-Hteik-Pan temple in Bagan. *Journal of Cultural Heritage* 43: 108–117. ISSN: 1296-2074. https://doi.org/10.1016/j.culher.2019.09.006. https://www.sciencedirect.com/science/article/pii/S129620741930010X.

De Luca, L. 2023. A digital ecosystem for the multidisciplinary study of Notre- Dame de Paris. *Journal of Cultural Heritage*. ISSN: 1269–2074. https://doi.org/10.1016/j.culher.2023.09.011. https://www.sciencedirect.com/science/article/pii/S129620742300184X.

Deschaux, J. 2017. Flood-related impacts on cultural heritage. *Floods* 1: 53–72. https://doi.org/10.1016/b978-1-78548-268-7.50004-3.

Ferracuti, B., et al. 2022. Damage to churches after the 2016 central Italy seismic sequence. *Geosciences* 12. https://doi.org/10.3390/geosciences12030122.

Google. 2020. 10 Heritage Sites Lost to Disaster and War. https://artsandculture.google.com/story/10-heritage-sites-lost-to-disaster-and-war/kALyuo79hhrkLQ. Accessed 06 Sept 2020.

Iizuka, A. 2021. Local performing arts and recovery from the Great East Japan earthquake and tsunami: A descriptive qualitative study. *International Journal of Disaster Risk Reduction* 63. ISSN: 2212-4209. https://doi.org/10.1016/j.ijdrr.2021.102446. https://www.sciencedirect.com/science/article/pii/S2212420921004076.

Jacquot, K., and R. Saleri. 2023. Gathering, integration, and interpretation of heterogeneous data for the virtual reconstruction of the Notre Dame de Paris roof structure. *Journal of Cultural Heritage*. ISSN: 1269–2074. https://doi.org/10.1016/jculher.2023.06.010. https://www.sciencedirect.com/science/article/pii/S1296207423001103.

Kallas, J., and R. Napolitano. 2023. Image-based 3D modeling as a damage prioritization tool for historic buildings in post-disaster areas: The case of the 2020 Beirut blast. *Journal of Cultural Heritage*, 314–321. ISSN: 1296-2074. https://doi.org/10.1016/j.culher.2023.06.007. https://www.sciencedirect.com/science/article/pii/S1296207423001061.

Lanza, S. 2003. Flood hazard threat on cultural heritage in the town of Genoa (Italy). *Journal of Cultural Heritage* 4 (3): 159–167. ISSN: 1296-2074. https://doi.org/10.1016/S1296-2074(03)00042-6. https://www.sciencedirect.com/science/article/pii/S1296207403000426.

Li, L., and Y. Tang. 2024. Towards the contemporary conservation of cultural heritages: An overview of their conservation history. *Heritage* 7 (1): 175–192. ISSN: 2571-9408. https://doi.org/10.3390/heritage7010009. https://www.mdpi.com/2571-9408/7/1/9.

Mezzino, D., et al. 2017. Built heritage documentation and management: an integrated conservation approach in Bagan. *ISPRS Annals of the Photogrammetry, Remote Sensing and Spatial Information Sciences* IV-2/W2: 143–150. https://doi.org/10.5194/isprs-annals-IV-2-W2-143-2017.

Ministry of and Culture, Religious Affairs. 2017. Bagan Disaster Risk Management Plan. https://www.gfdrr.org/sites/default/files/publication/Bagan%20DRM%20Eng.pdf.

Neumayer, E. 2004. The impact of political violence on tourism: Dynamic cross-national estimation. *Journal of Conflict Resolution* 48 (2): 259–281. https://doi.org/10.1177/0022002703262358.

Neumayer, E., and T. Plumper. 2016. Spatial spill-overs from terrorism on tourism: Western victims in Islamic destination countries. *Public Choice* 169: 195–206. https://doi.org/10.1007/s11127-016-0359-y.

Okamura, K. 2015. Tsunami and heritage after the 2011 Great East Japan Earthquake. In *Water and heritage: Material, conceptual and spiritual connections*, ed. W.J.H. Willems and H.P.J. van Schaik, 245–255. Leiden: Sidestone Press. ISBN: 978-90-8890-278-9.

Pavlova, I., et al. 2017. Global overview of the geological hazard exposure and disaster risk awareness at world heritage sites. *Journal of Cultural Heritage* 28: 151–157. ISSN: 1296-2074. https://doi.org/10.1016/j.culher.2015.11.001. https://www.sciencedirect.com/science/article/pii/S129620742300184X.

Turku, H. 2018. *The destruction of cultural property as a weapon of war. ISIS in Syria and Iraq*, 1st ed., 207. Palgrave Macmillan. ISBN: 978-3-319-57281-9. https://doi.org/10.1007/978-3-319-57282-6.

UNESCO. 1954. 1954 Convention for the Protection of Cultural Property in the Event of Armed Conflict. https://en.unesco.org/protecting-heritage/convention-and-protocols/1954-convention. Accessed 05 Jan 2024.

UNESCO. 1991. Paris, Banks of the Seine. https://whc.unesco.org/en/list/600. Accessed 06 Sept 2020.

UNESCO. 2019. Bagan. https://whc.unesco.org/en/list/1588. Accessed 06 Sept 2020.

UNESCO. 2025. Damaged cultural sites in Ukraine verified by UNESCO. https://www.unesco.org/en/articles/damaged-cultural-sites-ukraine-verified-unesco. Accessed 01 Mar 2025.

United Nations Office for Disaster Risk Reduction. 2015. Sendai framework for disaster risk reduction 2015–2030. https://www.preventionweb.net/files/43291_sendaiframeworkfordrren.pdf. Accessed 02 Jan 2024.

Weise, K., D. Gautam, and H. Rodrigues. 2018. Response and rehabilitation of historic monuments after the Gorkha earthquake, 65–94. https://doi.org/10.1016/B978-0-12-812808-4.00004-3.

WMF. 2012. East Japan Earthquake Heritage Sites. 2012 and 2014 World Monuments Watch. https://www.wmf.org/project/east-japan-earthquake-heritage-sites. Accessed 06 Sept 2020.

Chapter 2
Recording and Documenting Cultural Heritage

Abstract This chapter delves into the intricate process of preserving cultural heritage through various recording and documentation methods. It begins with defining culture and heritage, emphasizing their significance in providing identity and promoting social cohesion. The chapter then explores the classification and application of topographic, image-based, range-based, and integrated methods for recording and documenting cultural heritage. Each method's strengths and limitations are discussed, highlighting the need for a multidisciplinary approach to capture the full essence of cultural objects and sites. The integration of these methods, such as combining laser scanning with photogrammetry, is presented as the most effective way to achieve comprehensive and accurate documentation, ensuring the preservation of both tangible and intangible cultural heritage for future generations.

2.1 What is Culture?

Culture is the characteristics and knowledge of a particular group of people. In fact, it encompasses social behavior and rules found in human societies where people live and act, as well as arts, beliefs, customs, habits, laws, and practices of the people as part of a community.

It can also be defined as shared norms of behavior and interactions, as well as constructs relating to cognition and understanding that are learned through socialization. Culture can be defined as the development of a group identity through social patterns that are unique to that group, and it is important to avoid making subjective judgments unless they are clearly labeled as such.

Many countries around the world are renowned for their rich cultures, with customs, traditions, art, music, and cuisine serving as continuous attractions for tourists.

According to the Cambridge Dictionary "*Culture is the way of life, especially the general customs and beliefs, of a particular group of people at a particular time*".

E. Stylianidis, *Exploring the Ethical Dimension in Recording and Documenting Cultural Heritage*, https://doi.org/10.1007/978-3-031-80034-4_2

The term *Western culture* was introduced to define the culture of European countries and those heavily influenced by European immigration. Its roots can be traced back to the Classical Period of the Greco-Roman era, specifically in the 4th and 5th centuries B.C.

Eastern culture, on the other hand, usually refers to the social norms of the countries of Asia. It was heavily influenced by religion as it developed and tended to be less formalized. In general, there is less of a distinction between the cosmic society and the religious philosophy in Eastern culture than in Western culture. However, the term is used to encompass a wide range of traditions and histories.

Culture is an important part of human societies, giving individuals and communities an understanding of their history and heritage and providing them with an identity and sense of belonging. It helps to bridge differences and promote understanding by facilitating communication and connection between people through shared languages and traditions. Culture is not only the source of inspiration and creativity for people and communities but also an engine of social cohesion. It can also have an economic value, supporting tourism and the creation of jobs in the arts and the creative industries.

(Raeff et al. 2020) explore the definition of culture and how it can be studied to gain insight into the processes that influence behaviour and development. The authors argue that the answers to these questions are dependent on how culture is defined or conceptualised. Therefore, to analyze culture and understand its significance and impact on development, it is essential to first conceptualize it. The conceptualization of culture is not an easy task. Like many other complex human phenomena, culture has been the subject of numerous definitions and interpretations.

"*What Is Culture? What Does It Do? What Should It Do?*". This is the title of (Johnson 2013) work. In this study, the author argues that the term culture should replace cultures due to the complexity of cultural experiences. Furthermore, culture should serve specific purposes, and the culture relevant to political discussions should shape fundamental institutions based on three core values. Ultimately, a normative functionalist perspective is defended, asserting that culture should serve particular ends.

2.2 What is Heritage?

The term *heritage* is usually used to refer to the inheritance or bequest of something from the past. It is a property that is or has the potential to be an heirloom and is usually of significant value. A prime example of this is a historic building that has been passed down from generation to generation. The term heritage is associated with objects of historical or cultural significance that are worthy of preservation. Heritage is a valuable asset that can be passed from one generation to the next, can be preserved or inherited, and has significant historical and cultural value.

It is easy to think of heritage as a physical *object*: for example, a building or a place that can be *owned* by someone and *passed on* to someone else. In addition to these physical objects and places of heritage, other things can be considered, such as heritage practices that are preserved or passed on from one generation to the next. Another powerful example is language, an essential characteristic of individuals and communities that defines identity. It is learned and passed down from adults to children, from generation to generation. Intangible aspects of heritage, such as language, literature, and traditional attire, play a crucial role in shaping our identity and helping us understand who we are.

Another key aspect of heritage is the methods and practices used to preserve it, regardless of its form. This is reflected in the decisions we make about what to retain and what to discard from the past, which memories to preserve and which to forget, and which historic buildings to save or replace with new ones. Heritage encompasses practices—habits and customs that define who we are as individuals and communities. These practices contribute to the construction of our collective social memory. We use both heritage objects and heritage practices to shape our understanding of the past, present, and future.

Heritage is also very important to humanity. Through heritage, people can understand their roots and history, both as individuals and as communities. It helps them to have a sense of identity and to feel a sense of belonging. Heritage is a valuable source of education. It offers a deep perception of past generations to understand the entire context that built our world. The economic and social dimensions of cultural heritage are also important. The contribution to job creation and the tourism industry, but also the promotion of social cohesion by bringing people together.

According to (Waterton and Watson 2015), heritage, which encompasses objects, representations, events, and memories, has been at the center of academic discourse and public policy in recent decades. Understandings of heritage have evolved to emphasize its cultural, political, and identity-related dimensions, with an initial focus on the classification and preservation of objects. More recently, a complex interplay of individual, affective, and experiential issues has led to a shift toward the study of engagement processes and the construction of meaning. As a result, traditional notions of heritage, including concepts such as "*authenticity*" and "*memory*", are being reevaluated and revitalized through interdisciplinary perspectives. This has led to new theoretical inquiries and debates within the field. This evolving discourse is challenging the assumptions of the past and underscoring the dynamic nature of heritage studies in contemporary society.

2.3 What is Cultural Heritage?

Cultural heritage reflects the diversity of civilizations, the history of humanity, and the peaceful co-existence of people throughout the world. It is inherited and transmitted by individuals and communities. It consists of the tangible artifacts and intangible attributes of a particular group or society that are inherited from past generations, preserved in the present, and passed on for the benefit of future generations. It includes tangible culture such as historic buildings, monuments, landscapes, books, works of art, paintings, artifacts, and others. However, cultural heritage also includes the intangible culture that has its origin in the traditions or living expressions that have been inherited from our ancestors and passed on to our descendants. Indicative examples include oral traditions, the performing arts, social practices, rituals, and festivals, among others. No matter what form they take, cultural heritage requires dedicated and vigorous efforts to preserve it. While some aspects may be significant due to their present or potential economic value, others hold importance because they evoke a sense of self-confidence and deep emotional connection. Perhaps it is because they make us, or our community, feel connected to something emotionally larger—whether to a country, a tradition, or a way of life.

However, according to UNESCO (UNESCO 2017), the term cultural heritage includes several main categories of heritage (Fig. 2.1):

- Cultural heritage
  - Tangible cultural heritage
    - movable cultural heritage, including coins, paintings, etc.
    - immovable cultural heritage, including archaeological sites, monuments, etc.
    - underwater cultural heritage, including shipwrecks, underwater ruins, etc.
  - Intangible cultural heritage: such as oral traditions, performing arts, rituals, etc.
- Natural heritage: natural sites with cultural significance, such as cultural landscapes, physical, biological or geological formations
- Heritage in the event of armed conflict.

Tangible cultural heritage preserves the physical evidence of a society's history and achievements, providing a concrete link to the past. Intangible heritage, on the other hand, fosters community identity and continuity, enriching cultural diversity

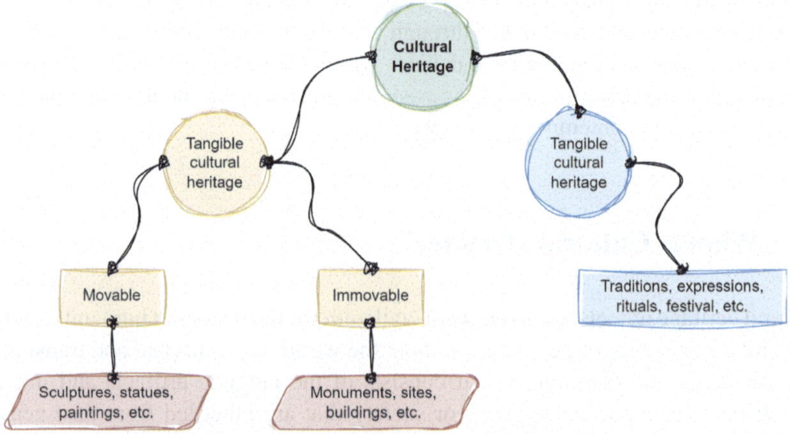

Fig. 2.1 Cultural heritage encompasses several main categories of heritage

and ensuring the transmission of knowledge and values across generations. Both forms of cultural heritage (Fig. 2.2) are essential for understanding and appreciating the full spectrum of human cultural expression.

In their work, (Vlase and Lahdesmaki 2023), the authors investigate the relationship between the evolution of cultural heritage research and the use of the Web of Science (WOS) as a knowledge management tool in academia. By analyzing nearly 2,000 journal articles published between 2003 and 2022 and indexed in WOS, this study reveals a Eurocentric bias in cultural heritage research, reflecting the dominance of European funding agencies. Despite a significant increase in published research, the concentration of studies within a few European institutions and countries suggests limited global collaboration. Themes such as digital technology and participatory approaches emerge prominently, indicating shifts in heritage care practices. Through bibliometric analysis and visualization using VOSviewer software, the authors shed light on the dynamics of research collaboration and knowledge production in the cultural heritage domain. They critically evaluate the role of WOS in capturing and managing scholarly output, offering insights into the broader context of international heritage governance and scientific production.

(Willis 2014) notes that cultural heritage preservation includes not only the tangible artifacts housed in museums and galleries but also the architectural marvels and historic sites protected by governmental regulations. These artifacts and places are often considered invaluable, raising economic questions about the investment required for their conservation. The author emphasizes the importance of investigating society's willingness to invest in preserving heritage for future generations, beyond just considering consumer demand for cultural experiences. This involves determining whether greater value is placed on the collective experience of historic areas rather than on isolated preservation efforts. Furthermore, the concepts of physical and social distance add complexity to individuals' preferences and the perceived value of cultural assets, necessitating further examination in economic evaluations of heritage conservation.

2.4 Recording and Documentation: the Profile of Cultural Heritage

Each cultural heritage object possesses unique characteristics. Based on the object's physical condition, time constraints, and required level of accuracy, a multidisciplinary team is responsible for determining the appropriate documentation workflow. It is important to consider the applicability of the established requirements. A classification of the tools and techniques available can be helpful in understanding and managing them effectively. Given the variety of techniques used in the field of cultural heritage documentation, multiple classification systems exist. One of the most common categorizations is based on metric data capture. The four main categories analyzed in this section are illustrated in Fig. 2.3.

(a)

(b)

Fig. 2.2 Tangible cultural heritage: Acropolis in Athens, Greece 2.2a; Intangible cultural heritage: folklore and dance 2.2b. (*Source* Pixabay)

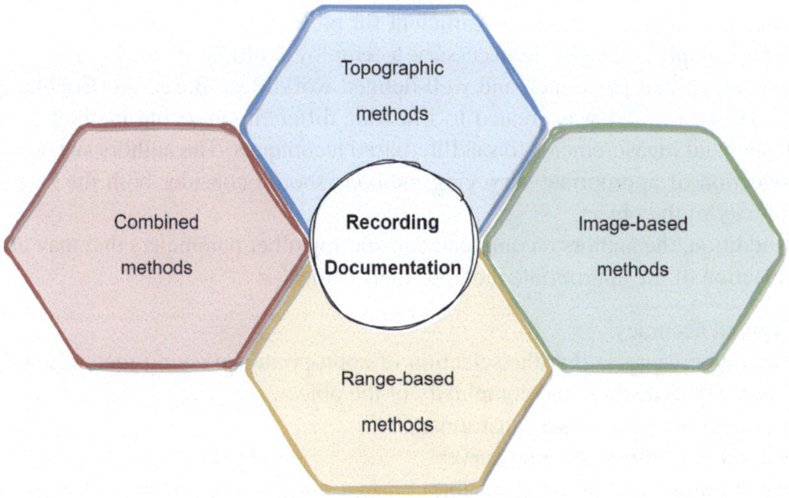

Fig. 2.3 Recording and documentation methods

- Topographic methods
- Image-based methods
- Range-based methods
- Integrated methods.

A team of experts from various disciplines selects the appropriate capture and documentation approach or tool based on the unique attributes of each object. Factors such as physical condition, time constraints, required accuracy, and project specifications are carefully. To enhance understanding and management, experts propose various classification systems for these techniques. The categorization presented here is based on capturing metric data and object point coordinates, with or without the use of images. The following sections explore these four main categories in detail.

The use of various technologies to record and document cultural heritage objects highlights the wide range of options available for collecting object metrics. It is important to note, however, that no single method can quarantee absolute accuracy, as each technique has its own set of limitations and challenges. Factors such as cost, time, complexity, accessibility, and the expertise of the survey team influence the choice of survey method. Each technique has unique characteristics. While each technique has unique characteristics, relying on a single approach is often insufficient to meet project requirements. Instead, a combination of techniques is usually necessary to achieve the desired results. When feasible within the project's budget, this hybrid approach is considered the most effective method for cultural heritage documentation.

In practice, there are a number of 2D and 3D surveying methods and techniques that can be used for cultural heritage recording and documentation. Simple tools, equipment, and methods may be sufficient for projects with low complexity. However, for complex cases, it is necessary to employ multiple methods, appropriate equipment, skilled personnel, and well-defined workflows. Based on (Boehler and Heinz 1999), Fig. 2.4 was created to illustrate different surveying methods, ranging from hand measurements to satellite-based techniques. The authors suggest that the selection of appropriate surveying methods should consider both the size and complexity of the object.

In addition, the authors recommend considering other parameters that may affect the selection of the appropriate method, such as the:

- Required accuracy.
- The authors suggest that the selection of appropriate surveying methods should consider both the size and complexity of the object.
- Unavailability of necessary instruments.
- Accessibility of object under survey.
- Availability of ideally located vibration-free observation and measurement stations.
- Authorization to touch an object under survey.

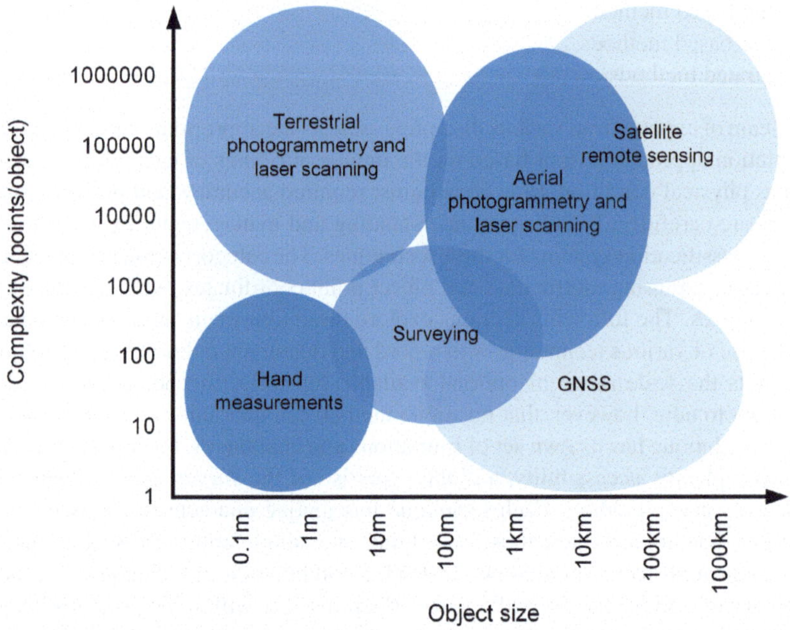

Fig. 2.4 2D/3D surveying techniques defined by object complexity and size

2.4.1 Topographic Methods

Topographic methods are critical to the accurate recording and documentation of cultural heritage sites through meticulous surveying techniques. Topography is the science of measuring the Earth's surface and its features. It involves the creation of detailed maps, charts, and plans to represent both natural and man-made elements. Natural features, such as mountains, rivers, valleys, and forests, are precisely mapped alongside human-made structures like buildings, roads, bridges, and settlements. Surveys depict slopes using contour lines or shading techniques to account for the slope of the land, which is critical to understanding terrain characteristics and drainage patterns. Topography aims to illustrate the spatial context and relationships of cultural resources within the landscape, encompassing a broader scope than surveying, which focuses on precise measurements of distances, angles, and elevations.

Topographic surveys are essential for conserving and managing cultural heritage sites. They help researchers and professionals make informed decisions about preservation and development by providing detailed documentation of site layouts, dimensions, and spatial configurations. Accurate depiction of the physical and cultural features of the landscape through topographic mapping aids in the identification, protection, and interpretation of heritage resources. Surveys facilitate proactive conservation strategies by allowing monitoring of environmental changes such as erosion and land use impacts. Topographic methods are critical to preserving cultural heritage for future generations. They ensure comprehensive documentation and informed management practices.

Topographic measurements provide a direct method for collecting spatial information, resulting in discrete coordinates of specific points. These points can be used to create topographical maps, plans, and sections. The captured spatial information serves to:

- Measure distances, angles, slopes, and differences in height between points.
- Set out straight lines, perpendiculars, and parallels in the field of measurements.
- Determine horizontal and vertical lines.
- Calculate areas and volumes.

In practice, a sufficient number of detail points with 3D coordinates must be recorded for accurate geometric documentation. The choice of method depends on the required accuracy of the final product, the available equipment, and the time allocated for fieldwork. Two methods can be used: the *forward intersection method* (FIM) or the *method of polar coordinates* (MPC).

The FIM method for determining detail points in 3D space has gradually fallen out of favor in recent years due to advancements in technology and the emergence of more efficient alternatives. The FIM traditionally relied on a pair of topographic instruments, typically two theodolites or total stations, along with additional equipment such as tripods and a tape measure to measure the heights of the instruments.

The FIM process uses two known coordinates, A and B, as reference points to determine the coordinates of the target point, M. Visibility between A, B, and M is necessary. Measurements collected include horizontal and vertical angles from both A and B, as well as the heights of the instruments. These data are then used in the FIM to calculate the coordinates of point M with reasonable accuracy, typically within a few centimeters. However, despite its historical significance, the FIM method has largely been superseded in many applications by the adoption of reflectorless total stations. These stations offer greater convenience, speed, and precision in capturing spatial data eliminating the need for physical reflectors or extensive setup requirements.

MPC is a method used in geodetic surveying to determine the 3D coordinates of specific points. The process is generally referred to as the use of total stations. In geodetic and other forms of surveying, the total station is the most common method for determining the 3D coordinates of detail points. This method is particularly used when distances to the points of interest can be measured directly. The instruments used are typically total stations. These instruments combine electronic distance measurement and electronic angle measurement to calculate precise positions. They can be used with a reflector or without a reflector. In the case where a reflector is used, a small, highly reflective target is placed on top of the point that needs to be surveyed. The total station sends a laser beam to the reflector and the beam bounces back to the total station. The instrument then calculates the distance on the basis of the time taken for the return of the light. Reflectorless mode refers to the use of total stations that measure distances without requiring a physical reflector at the target point. This is useful in cases where it is not possible to place a reflector on the target. The station emits a laser. The laser measures distance directly by detecting the light reflected from the surface of the target. A typical workflow of the MPC method is as follows:

- Set up the total station at point A.
- Level the total station.
- Measure total station height.
- Target at point B.
- Initialize the azimuth at point B.
- Position reflector system vertically at point C.
- Target at point C.
- Measure and record the horizontal and vertical angles, the distance, and target height.

Topographic measurements are used to determine the coordinates of discrete points, which can then be utilized to create detailed designs or plans. Topographic surveying is crucial for accurately calculating the coordinates of discrete points on a landscape or structure. These coordinates provide a precise framework that can be used to create detailed designs or representations. For example, using a series of such points in specific positions allows the reconstruction or depiction of complex architectural designs as demonstrated in the referenced figures. For example, Fig. 2.5a showcases the floor plan of a 9th-century Byzantine church. This representation is derived from topographic measurements that define the spatial relationships and dimensions within the church, offering insights into the architectural style and spatial dynamics

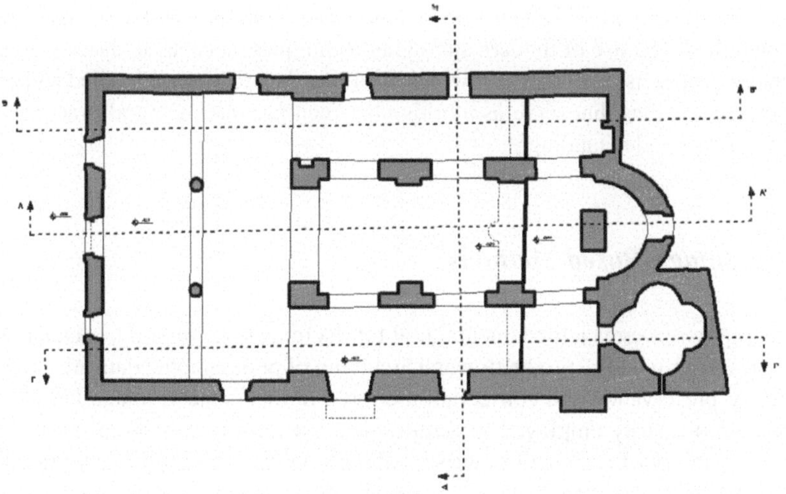

(a)

(b)

Fig. 2.5 A ground plan 2.5a and a cross-section 2.5b can be designed using topographic measurements

of the period. Similarly, Fig. 2.5b illustrates a cross-section of the church, providing a graphical representation derived from topographic data. This cross-section provides a vertical profile of the structure, revealing the heights and structural depths that are not visible in the floor plan.

These topographic representations are invaluable for historical and archaeological studies, allowing researchers to visualize and analyze the architectural elements and construction techniques of ancient structures without physically altering the site. They also enable the accurate restoration of heritage buildings, ensuring that conservation respect the original design and structural integrity. Additionally, such detailed drawings are crucial for educational purposes, helping students understand the architectural evolution and historical context of significant buildings like Byzantine churches. The use of modern surveying techniques, such as 3D laser scanning, enhances the precision of these drawings, enabling the creation of detailed and accurate representations that serve as a bridge between past architectural practices and contemporary understanding.

2.4.2 Image-Based Methods

Photography serves as a fundamental tool for documenting cultural heritage at specific points in time. The production and archiving of photographic data are essential for future preservation and conservation efforts. In this context, image-based techniques are primarily employed to acquire detailed information about the targeted objects, with coordinates and other spatial data typically derived during the post-processing phase. Among these techniques, photogrammetry and the use of infrared (IR) cameras are particularly significant.

Image-Based Modeling (IBM) is a technique that utilizes images or videos to create three-dimensional models of objects or scenes. This process involves extracting information about the shape and texture of the objects from the images, and then using this information to construct accurate 3D models.

There are a number of techniques that can be used for IBM, such as the following:

Photogrammetry: involves creating 3D models from photographs taken from multiple angles. By measuring and analyzing these images, one can calculate the 3D coordinates needed to build detailed models of objects or scenes. It is widely used in archaeology to reconstruct sites or in architecture to create models of historic buildings for restoration purposes (Fig. 2.6).

Structure from Motion (SfM): uses sequences of images captured from different viewpoints. It detects common features in these images to determine the camera's position and orientation at each shot, enabling the construction of a 3D model. This method is particularly useful for creating 3D models of large-scale landscapes and complex structures where traditional surveying is challenging or invasive.

Multi-View Stereo (MVS): extends the principles of SfM by using a broader array of images to generate more comprehensive and detailed 3D models. It compares features across multiple images to build depth maps that contribute to the model's accuracy. It is used in detailed documentation of intricate objects, providing high-resolution models essential for detailed analysis and preservation.

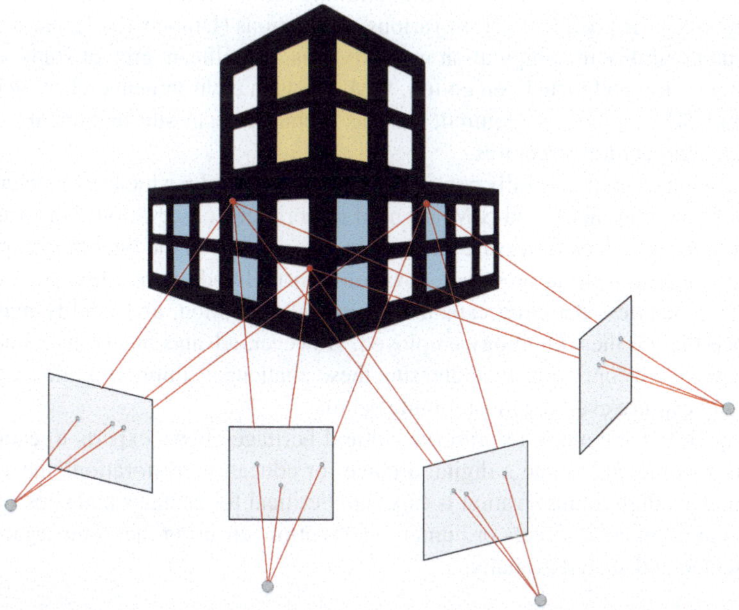

Fig. 2.6 Photogrammetry is the process of extracting accurate measurements from multiple overlapping images

The use of these IBM techniques in cultural heritage conservation is invaluable. They allow for the non-invasive documentation of artifacts, buildings, and sites, which is crucial for preserving their integrity while still allowing for detailed studies. These models can serve various purposes, including virtual tours, condition assessment, restoration planning, and inclusion in digital archives for future generations.

IBM is extensively utilized across various fields, including the recording and documentation of cultural heritage. However, this technology is has its limitations. For effective results, IBM requires optimal lighting conditions and distinct textures. It often struggles with capturing complex geometries or transparent and reflective objects. Additionally, the accuracy of the generated model heavily depends on the quality of the images used and the capabilities of the processing software.

According to the Statutes of the International Society for Photogrammetry and Remote Sensing (ISPRS), "*Photogrammetry is the science and technology of extracting reliable three-dimensional geometric and thematic information, often over time, of objects and scenes from image and range data.*" (ISPRS 2022) This definition underscores the methodological essence of photogrammetry, which involves taking multiple overlapping images of an object. These images are then processed with sophisticated algorithms to produce a three-dimensional model. The key principle of photogrammetry is to harness the multiple perspectives offered by these images to accurately reconstruct the spatial characteristics and details of the depicted objects.

SfM is another vital process that estimates the 3D structure of a scene using a set of images. SfM is prevalent across various applications (Ullman 1979), particularly due to its flexibility in computation methods. It is a significant area of study within computer vision and visual perception. SfM excels in environments where detailed and rapid 3D mapping is required, such as archaeological site documentation or complex architectural structures.

Professionals must carefully consider the conditions under which IBM techniques such as photogrammetry and SfM are used in heritage conservation. For example, uneven lighting or lack of texture can lead to inaccuracies in the final model, potentially misguiding restoration or conservation efforts. In addition, while IBM offers valuable contactless detailing capabilities, model resolution, and fidelity may differ depending on the technology employed. However, advancements in technology and software are continuously addressing these challenges, improving the ability to accurately capture even the most minute details.

Using IBM technology to preserve cultural heritage allows experts to capture a moment of time and create a digital archive for education, restoration, and virtual tourism. This digital preservation is especially critical for artifacts and sites vulnerable to environmental threats or human intervention, ensuring that their legacy can be protected and shared globally.

2.4.3 Range-Based Methods

Range-Based Modeling (RBM) is a technique used to create three-dimensional models by measuring the distance between a sensor and the object of interest. These measurements are collected as a point cloud, which accurately represents the surface geometry of the object or scene.

There are a number of techniques for RBM, including the following:

Laser scanning: uses a laser to systematically scan an object or area. Each laser pulse measures the distance to a point on the object's surface, creating a dense point cloud that can be used to reconstruct a detailed 3D model. It is widely used in surveying, and industrial design, and it is value for the precision and the high resolution of the models it produces.

Light Detection and Ranging (LiDAR): uses a pulsed laser to measure distances, similar to laser scanning. However, it combines these measurements with positional data, i.e. from GNSS, to create large-scale, high-resolution models over large areas. It is particularly valuable in topography and forestry, and has important applications in autonomous vehicle technology, where accurate, real-time mapping of the environment is critical.

Time-of-Flight (ToF) cameras: measure the time it takes for light to travel to the object and back to the camera sensor. This method captures distance information at each pixel, allowing a 3D model to be created based on the timing of light reflected from the object. It is used in both industrial applications, such as robot navigation, and in consumer technology, including augmented reality and interactive gaming.

The applications of RBM are numerous and diverse, ranging from the documentation of cultural heritage to engineering, to virtual reality, among others. In the cultural heritage field, for example, RBM enables the accurate documentation and preservation artifacts and historic sites. In engineering, these models assist in planning and monitoring construction projects, as well as in the maintenance of critical infrastructure.

Despite its benefits, RBM also has limitations. Weather conditions may reduce the effectiveness of the system, as heavy rain, fog, or bright sun may obscure the laser pulses and light signal. In addition, the equipment can be expensive and requires specialized knowledge to operate effectively. A common solution to these challenges is the integration of RBM with other technologies, such as photogrammetry or advanced imaging software, to enhance the accuracy and utility of the data. This integrated approach leads to more precise measurements and detailed spatial analysis, ensuring that RBM remains a valuable tool across a wide range of fields.

Laser scanning is one of the predominant methods for capturing the physical properties of objects and environments, particularly in the field of cultural heritage. This technology, which is particularly useful for recording and documenting historical artifacts and sites, uses laser light to scan and capture high-resolution, three-dimensional data. There are several types of 3D laser scanners on the market, each differing in technical specifications, capabilities, and cost. Therefore, selecting the right scanner based on the specific needs of the project is crucial.

Capturing and documenting cultural heritage with 3D laser scanning requires careful planning and a deep understanding of the object and its historical context. The process begins with a thorough assessment of the site and the specific features to be documented. This preliminary phase is crucial for determining the appropriate scanning equipment and selecting the most suitable techniques. Once these factors are established, a detailed workflow (Fig. 2.7) is developed to guide the scanning process effectively.

Preparation and planning: The initial phase of the process requires meticulous preparation and planning. Scanning team carefully evaluates the object to be measured and identifies the critical areas of interest. To ensure comprehensive data collection, the team strategically selects optimal scanning positions. This phase involves a thorough assessment of factors such as lighting conditions, accessibility, visibility, and the implementation of safety measures for all team members. Careful consideration of these elements is crucial to ensuring the effectiveness and accuracy of subsequent scanning activities.

Scanning: Once the preparation and planning phase has been completed, the team initiates the actual 3D laser scanning process. The laser scanner is positioned at multiple vantage points, and by releasing the laser beam, it traverses the surfaces of the object to be scanned. As the laser makes contact with the surfaces, it measures distance and captures a large number of 3D data points. The process generates a dense point cloud, representing the 3D geometry of the object. To

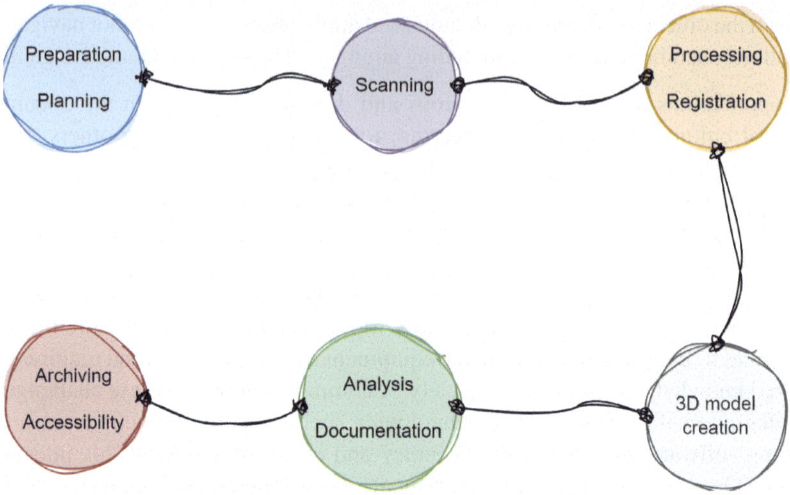

Fig. 2.7 3D laser scanning workflow

ensure thorough coverage and capture fine details, the team conducts multiple scans from various angles and locations, optimizing the data-collection process. This approach ensures a comprehensive and accurate representation of the object.

Processing and registration: After completing a 3D laser scan of an object, the acquired data must be processed and registered to create a cohesive and accurate representation of the object. To eliminate any inconsistencies or misalignments, specialized software is used to align and consolidate the individual scans. A seamless and accurate 3D model of the entire object is created through this registration process. The high level of accuracy achieved through this registration process enhances the overall quality of the 3D model, making it a reliable tool for subsequent analysis, documentation and immersive exploration.

3D model creation: After the laser scan data is successfully registered, the point cloud is transformed into a detailed 3D model. A digital representation of the object is created by processing the point cloud to reconstruct surfaces. This digital model provides a 360° virtual tour of the object, allowing it to be viewed and explored from multiple perspectives. Experts can use this immersive tool for in-depth analysis. The general public benefits from the ability to virtually navigate and explore the object. Fostering a broader understanding of the historical and architectural significance of the object, this virtual tour serves as an accessible and engaging means for diverse audiences to connect with and appreciate the intricacies of the object.

Analysis and documentation: The 3D model created through laser scanning is a critical asset for in-depth analysis and comprehensive documentation. Professionals benefit greatly from the ability to examine the model closely, explore intricate architectural features, and accurately identify areas of deterioration or damage. This detailed examination enables precise measurements to be taken directly from the model, significantly enhancing meticulous restoration planning, targeted conservation efforts, and the advancement of historical research initiatives. As a versatile tool, 3D modeling not only aids in understanding the nuances of the object but also improves decision-making processes across various fields, promoting informed approaches to architectural preservation and historical research.

Archiving and accessibility: Upon completion of the analysis and documentation, the 3D digital model is carefully archived to ensure long-term preservation and accessibility. This practice safeguards the preservation of the valuable information obtained through the 3D laser scanning of cultural heritage objects, making it available to future generations for research, education, and conservation efforts. The archived model serves as a comprehensive resource for both internal use and sharing with organizations, museums, and public audiences. This sharing enables diverse audiences to engage with and learn from the intricate details of cultural heritage objects, fostering a broader understanding and appreciation of our architectural heritage. The accessibility of this archived data plays a critical role in the promotion of cultural heritage awareness and the fostering of collaborative efforts to preserve and celebrate our architectural legacy.

The primary goal of a laser scanning campaign is to accurately capture the intricate architectural details and overall structure of an object. The following figures (Fig. 2.8) illustrate an example of such a campaign conducted at the Church of the Savior in the city of Thessaloniki, Greece. The church is a Byzantine monument built around 1350 (14th century). It was inscribed on the UNESCO World Heritage List in 1988. The church was severely damaged by a massive earthquake in 1978. The non-intrusive method is a guarantee of high accuracy in the capture of the unique features of the church. The resulting 3D point cloud facilitates the creation of an accurate digital twin, providing invaluable data to architects, historians, and restoration specialists for informed preservation and restoration decisions.

Multiple scanning sessions were scheduled during the campaign to ensure complete coverage of the church. The collected data will be processed using state-of-the-art software. The individual scans will be stitched together to create a cohesive, highly detailed 3D model. This model will significantly aid ongoing conservation efforts and will also enhance public engagement and educational projects by enabling virtual tours and facilitating detailed analysis of the church's architectural heritage.

2.4.4 Integrated Methods

In recent years, emerging technologies have become essential tools for cultural heritage professionals in analyzing, processing, reconstructing, interpreting, managing, and disseminating data related to various cultural heritage objects.

The field of Geomatics offers numerous solutions for comprehensive digital metric surveying, recording and documentation of cultural heritage. Captured data can

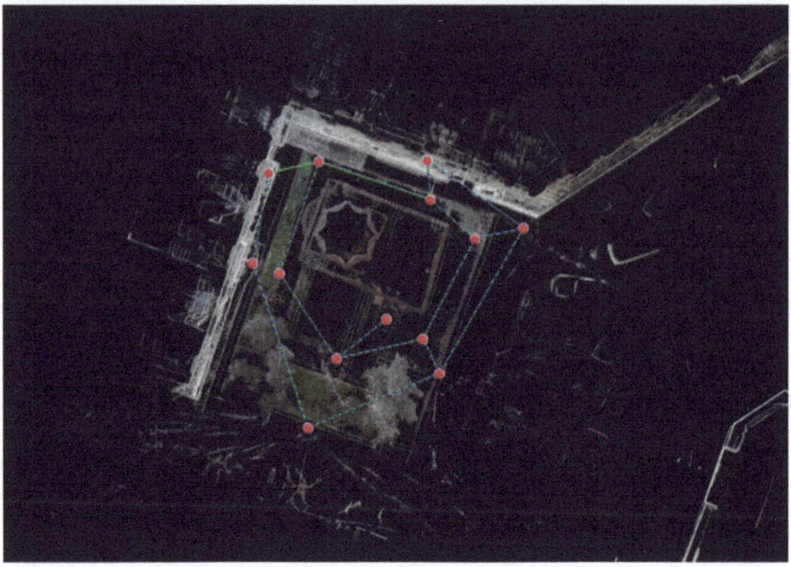

(a)

(b)

Fig. 2.8 Plan view of the laser scanning campaign for the Church of the Savior in the city of Thessaloniki, Greece 2.8a; the captured 3D point cloud 2.8b

be processed to generate precise metric information that is useful not only for documentation, but also for conservation purposes. Additionally, this information can be effectively shared through databases or GIS tools, enhancing accessibility and supporting informed decision-making in heritage management.

For the geometric and architectural documentation of cultural heritage objects, structures and sites, a combination of survey methods can be employed. In scenarios where traditional survey techniques are insufficient, or when time constraints and high accuracy requirements are critical, alternative methods are often necessary. For example, close-range aerial digital photogrammetry—augmented by Global Navigation Satellite System (GNSS), total station positioning systems, and laser scanners—can be used at various levels of detail to produce detailed 2D drawings, maps, and 3D models.

The use of combined and integrated methods enables comprehensive and detailed documentation, including heterogeneous information about the cultural heritage object. This includes not only the architectural and geometric characteristics of the object, but also its archaeological and historical elements, construction materials, stylistic features, and evidence of deterioration.

Combined methods for recording and documenting cultural heritage involve the integration of various techniques, including topographic, image-based, and range-based modeling. These approaches are used to create comprehensive 3D representations of cultural heritage monuments, sites, artifacts, or traditions.

Integrating these methods increases both the accuracy and richness of the collected data. Topographic modeling provides a detailed representation of the physical features and terrain of a heritage site. Image-based techniques involve the acquisition of high-resolution imagery, enabling the visualization of intricate details and surface textures. Range-based modeling employs technologies such as laser scanning or LiDAR to measure distances precisely, generating accurate 3D point clouds.

By combining these methods, practitioners can create holistic and detailed documentation of cultural heritage. This approach not only captures the visual aspects but also preserves the spatial relationships and physical characteristics of the recorded elements. The resulting 3D representation serves as a valuable resource for conservation, research, and educational purposes, providing a virtual window into the past that can be studied and shared with a wide range of audiences. In addition, the use of integrated methods enables a more nuanced understanding of cultural heritage by capturing details at both macro and micro levels, ensuring an accurate and complete representation of the subject matter.

Integrated methods play a critical role in overcoming the inherent limitations of individual imaging techniques. For example, while distance-based methods excel at capturing precise geometric data, they often struggle to accurately reproduce texture and color details. Conversely, image-based methods excel at capturing rich visual information but may encounter difficulties when documenting intricate or transparent objects.

By integrating range-based and image-based methods, these limitations can be effectively mitigated. Range-based data deliver precise geometric information, ensuring accurate representation of the structural aspects of heritage monuments, sites, or artifacts. Meanwhile, image-based data enhance visual fidelity by capturing intricate surface details, textures, and colors that would otherwise be difficult to capture with range-based methods alone.

This synergy results in a more complete representation of the cultural heritage that addresses the shortcomings of each method. The combined dataset not only provides a precise representation of physical characteristics, but also a richer, more immersive experience for researchers, conservators, and the general public. Such an integrated approach is especially valuable for addressing the complexities and nuances associated with diverse cultural heritage elements, including monuments, artifacts, sites, and traditions.

An exemplary integration of methods is the combination of terrestrial laser scanning (TLS) with photogrammetry. In this approach, laser scanning captures the precise geometric details of an object or scene, while photogrammetry captures texture and color information. Another common integration involves combining TLS, a total station, and a drone (Fig. 2.9).

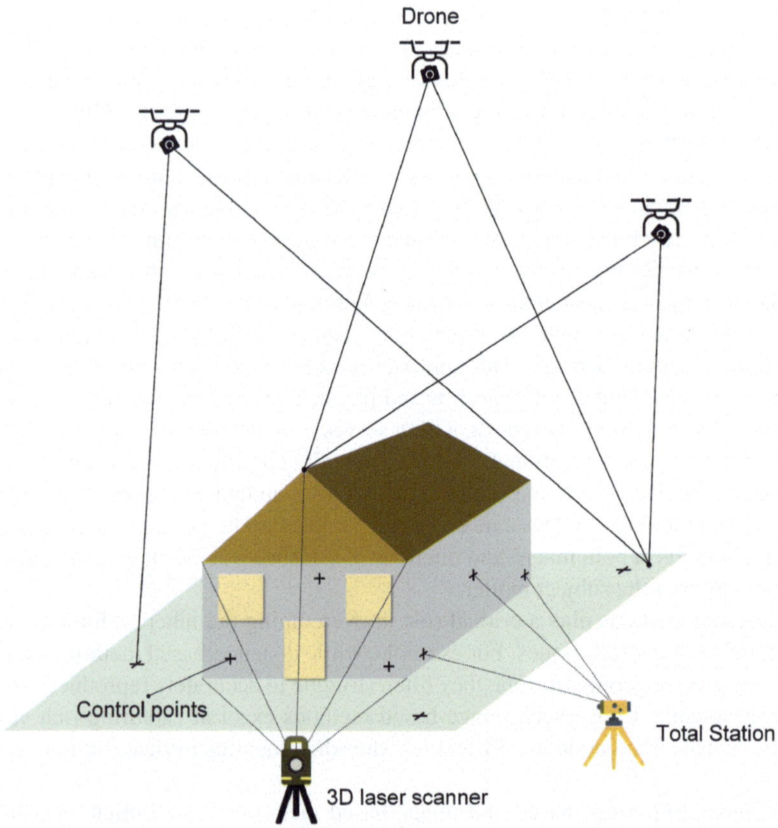

Fig. 2.9 A typical example of combined methods includes TLS, a total station, and a drone

Integrated methods for recording and documenting cultural heritage are extensively applied across diverse fields, including archaeology and architectural heritage. These integrated approaches, encompassing techniques such as topographic, image-based, and range-based modeling, yield outcomes that hold significant value for research, preservation, and education initiatives. Leveraging these comprehensive 3D representations promotes deeper understanding and appreciation of the cultural heritage of communities and regions.

In archaeological contexts, the utilization of combined methods facilitates the meticulous capture of excavation sites, artifacts, and surrounding landscapes. The detailed 3D models generated through these techniques enable more accurate analyses of historical contexts, empowering researchers to unravel the complexities of past civilizations.

In the realm of architectural heritage, these methods prove invaluable for the documentation and preservation of historical structures. The integration of topographic, image-based, and range-based data ensures a comprehensive representation, encompassing both the structural intricacies and aesthetic details of buildings. This extensive dataset serves as a valuable resource for conservation efforts, supporting professionals in the restoration and maintenance of architectural treasures.

The outcomes of these recording methods extend beyond academic research, significantly enhancing educational initiatives by providing immersive and interactive experiences. Virtual reconstructions of cultural heritage sites and artifacts offer a unique opportunities for students and the general public to engage with history. Furthermore, the accessibility of these 3D representations facilitates broader dissemination, fostering cultural awareness and a deeper sense of connection.

In essence, the widespread application of integrated methods for recording and documenting of cultural heritage is pivotal to preserving and promoting the diverse legacies of communities and regions. This approach fosters a more informed and enriched understanding of our shared past.

References

Boehler, W., and G. Heinz. 1999. Documentation, surveying, photogrammetry. In *Proceedings 17th international symposium CIPA 1999*. Olinda, Brazil.

ISPRS. 2022. Statutes—international society for photogrammetry and remote sensing. https://www.isprs.org/documents/statutes.aspx. Accessed 28 Dec 2023.

Johnson, M. 2013. What is culture? what does it do? what should it do? In *Evaluating culture: well-being, institutions and circumstance*, 97–119. London: Palgrave Macmillan UK. ISBN: 978-1-137-31379-9. https://doi.org/10.1057/9781137313799_5.

Raeff, C., et al. 2020. The concept of culture: introduction to spotlight series on conceptualizing culture. *Applied Developmental Science* 24 (4): 295–298. https://doi.org/10.1080/10888691.2020.1789344.

Ullman, S. 1979. The interpretation of structure from motion. *Proceedings of the Royal Society of London. Series B, Biological Sciences* 203: 405–426.

Vlase, I., and T. Lahdesmaki. 2023. A bibliometric analysis of cultural heritage research in the humanities: The Web of Science as a tool of knowledge management. *Humanities and Social Sciences Communications* 10 (84): 295–298. https://doi.org/10.1057/s41599-023-01582-5.

Waterton, E., and S. Watson. 2015. Heritage as a focus of research: past, present and new directions. In *The Palgrave handbook of contemporary heritage research*, 1–17. London: Palgrave Macmillan UK. ISBN: 978-1-137-29356-5. https://doi.org/10.1057/9781137293565_1.

Waterton, E., and S. Watson. 2017. What is meant by "cultural heritage"? http://www.unesco.org/new/en/culture/themes/illicit-trafficking-of-cultural-property/unesco-database-of-national-cultural-heritage-laws/frequently-asked-questions/definition-of-the-cultural-heritage/. Accessed 26 Aug 2020.

Willis, K. 2014. Chapter 7—The use of stated preference methods to value cultural heritage. In *Handbook of the economics of art and culture*, vol. 2, ed. V. Ginsburgh, and D. Throsby, 145–181. Elsevier. https://doi.org/10.1016/B978-0-444-53776-8.00007-6. https://www.sciencedirect.com/science/article/pii/B9780444537768000076.

Chapter 3
The World Cultural Heritage Context

Abstract The chapter provides a comprehensive overview of the international framework for cultural heritage protection. It emphasizes the significance of preserving both tangible and intangible cultural heritage to foster a sense of belonging and historical continuity. Key international conventions are highlighted for protecting heritage sites and practices. The chapter also discusses the importance of regional agreements and national laws in reinforcing global norms and addresses contemporary challenges like climate change. Furthermore, it reviews the ICOMOS Principles for the Recording of Monuments and other key ethical standards guiding professionals in the conservation and management of cultural heritage. Through these frameworks, the chapter underscores the need for a coordinated, ethical approach to cultural heritage preservation, ensuring its resilience and relevance for future generations.

3.1 The International Framework

The international framework for the protection and management of cultural heritage is a complex and evolving structure that reflects the diverse values and priorities of communities worldwide. Cultural heritage fosters a sense of belonging by preserving the tangible and intangible legacies that connect people to their histories, traditions, and identities. This heritage monuments, buildings, and sites of historical, artistic, archaeological, scientific, ethnological, or anthropological significance. Preserving this heritage is essential to maintaining the cultural diversity and historical richness of humanity.

At the heart of this framework are international conventions and treaties under the leadership of organizations such as UNESCO (the United Nations Educational, Scientific and Cultural Organization). The 1972 World Heritage Convention 3.2, which aims to identify and protect the world's cultural and natural heritage of outstanding universal value, is one of the most important. The Convention has been ratified by nearly all UN member states, making it one of the most widely accepted legal instruments in this field. It emphasizes shared history through the protection of World Heritage sites, with member states pledging to safeguard their listed heritage sites and to refrain from any actions that could harm them.

E. Stylianidis, *Exploring the Ethical Dimension in Recording and Documenting Cultural Heritage*, https://doi.org/10.1007/978-3-031-80034-4_3

The focus on intangible cultural heritage, which includes traditions, performing arts, rituals, and other practices passed down through generations, is another important aspect of the international framework. The 2003 Convention for the Safeguarding of the Intangible Cultural Heritage outlines measures for documenting, preserving, and promoting these cultural forms, recognizing their importance for cultural identity and continuity. By ensuring that the dynamic nature of living cultural traditions is respected and that community members are involved in their preservation, it reinforces the idea that cultural heritage contributes to the resilience of our societies.

These conventions are complemented by numerous regional agreements and national laws that reinforce the global norms established by the universal framework. This multifaceted approach addresses the challenges posed by natural deterioration, human conflict, and globalization, and ensures that cultural heritage protection is globally coordinated and locally relevant. By protecting and promoting cultural heritage, these frameworks also contribute to sustainable development by helping societies balance economic growth with preserving their unique cultural assets. Emphasizing the importance of cultural heritage in building resilient and inclusive societies, the international community continues to adapt these frameworks to contemporary challenges such as climate change and digital reproduction.

3.2 World Heritage Convention of 1972

The protection of the world's cultural and natural heritage is a collective responsibility. According to the Convention *Concerning the Protection of the World Cultural and Natural Heritage* (UNESCO 1972), known as the *World Heritage Convention*, and specifically Article 6, the States Parties to this Convention recognize that such heritage constitutes a shared global legacy, whose protection is a duty of the international community as a whole to uphold through cooperation:

> *Whilst fully respecting the sovereignty of the States on whose territory the cultural and natural heritage mentioned in Aticles 1 and 2 is situated, and without prejudice to property right provided by national legislation, the States Parties to this Convention recognize that such heritage constitutes a world heritage for whose protection it is the duty of the international community as a whole to co-operate.*
>
> *The States Parties undertake, in accordance with the provisions of this Convention, to give their help in the identification, protection, conservation and presentation of the cultural and natural heritage referred to in paragraphs 2 and 4 of Article 11 if the States on whose territory it is situated so request.*
>
> *Each State Party to this Convention undertakes not to take any deliberate measures which might damage directly or indirectly the cultural and natural heritage referred to in Articles 1 and 2 situated on the territory of other States Parties to this Convention.*

The World Heritage Convention of 1972 (Fig. 3.1) integrates the concepts of cultural properties preservation and nature conservation in a single document approved by the international community. The Convention obligates states to identify potential sites and recognizes their distinct roles in protecting and preserving heritage. By

Fig. 3.1 Adoption of the World Heritage Convention in 1972 at UNESCO. (*Source* ©UNESCO)

signing the Convention, each country pledges to protect not only the World Heritage sites located within its territory but also its entire national heritage.

Although States identify potential sites, the World Heritage Committee (UNESCO 2020c) ultimately decides which sites will be inscribed on the World Heritage List. This list, published on UNESCO's official site (UNESCO 2020e), includes those sites with outstanding universal value, forming part of the world's cultural and natural heritage. Additionally, a Tentative List (UNESCO 2020b) serves as a provisional catalog of properties that a State intends to nominate for future inclusion on the World Heritage List.

The 1972 Convention Concerning the Protection of the World Cultural and Natural Heritage also establishes the List of World Heritage in Danger (UNESCO 2020a). The purpose of this list is to inform the international community about the threats faced by properties on the World Heritage List and to encourage the adoption of remedial measures for those at risk.

War and armed conflict pose a significant threat to World Heritage sites. Under the Convention, if specific circumstances make the protection of a property essential, the World Heritage Committee may inscribe it on the List of World Heritage in Danger. This process requires cooperation between the Committee and the concerned State to develop a restoration and monitoring program for the property. Once a property has been inscribed on the List of World Heritage in Danger, stakeholders work to restore its values and facilitate its removal from the list as soon as possible.

States are required to inform the Committee of any threats to their properties. In addition, individuals, non-governmental organizations, and other interested parties may bring potential threats to the attention of the Committee, provided that the concerns are justified. Inclusion on the List allows the Committee to utilize the World Heritage Fund (UNESCO 2020d) to provide direct assistance to the property under threat. In extreme cases, if a World Heritage property remains under threat and appropriate measures are not taken, it may lose its status as a World Heritage site.

3.3 ICOMOS Principles for the Recording of Monuments, Groups of Buildings and Sites of 1996

The ICOMOS document, *Principles for the Recording of Monuments, Groups of Buildings and Sites of 1996* (ICOMOS 2012), is a foundational reference for recording activities in cultural heritage documentation and preservation. This document extends Article 16 of the *International Charter for the Conservation and Restoration of Monuments and Sites*, commonly known as the *Venice Charter of 1964* (ICOMOS 1964), by broadening the scope of standards for recording heritage.

Article 16.
In all works of preservation, restoration or excavation, there should always be precise documentation in the form of analytical and critical reports, illustrated with drawings and photographs. Every stage of the work of clearing, consolidation, rearrangement and integration, as well as technical and formal features identified during the course of the work, should be included. This record should be placed in the archives of a public institution and made available to research workers. It is recommended that the report should be published.

The Charter articulates standards of practice for the conservation of monuments and sites, emphasizing the importance of thorough documentation. The ICOMOS document expands on this by defining comprehensive principles that consider the broader contexts in which heritage is defined and conserved. It provides a more inclusive framework than the monument-centric approaches envisioned by the Venice Charter over three decades earlier.

Ratified by the 11th ICOMOS General Assembly in Sofia, Bulgaria, in October 1996, the Principles for the Recording of Monuments, Groups of Buildings and Sites document is designed to be clear and accessible (Fig. 3.2). It consists of five distinct yet interconnected sections:

1. Introduction: this section explains the rationale behind recording cultural heritage, highlighting its significance and purpose.
2. Responsibility of recording: it outlines the importance of accountability in ensuring the quality and content of the records.
3. Planning procedures: this section details the essential planning processes for recording cultural heritage, emphasizing the need for thorough preparation.

Fig. 3.2 The cover of the ICOMOS 11th General Assembly, Sofia Bulgaria, 1996

4. Content of records: this section outlines the details of the information collected during the recording process, emphasizing the importance of comprehensive documentation.
5. Management, dissemination, and sharing: the final section focuses on managing the collected records and ensuring their availability and accessibility for broader use.

This document serves as a guide for practitioners and institutions responsible for cultural heritage documentation. It provides a consistent approach that emphasizes the importance of quality, integrity, and accessibility in the recording and documentation of cultural heritage.

3.3.1 Reasons for the Recording of Cultural Heritage

The first section of the ICOMOS document defines two key terms: *cultural heritage* and *records*. It then explores the importance of recording cultural heritage, emphasizing the need to adhere to specific standards. It recognizes the inherent value of cultural heritage and the various risks it faces, acknowledging that it is often vulnerable to natural and human-made threats. Recording is identified as a fundamental measure for safeguarding cultural heritage.

These conditions highlight the critical importance of recording and documentation. The responsibility for cultural heritage is shared among various stakeholders. Recording and documentation should be a collaborative effort involving owners, managers, professionals, and others engaged with cultural heritage.

This first section of the document reinforces the overarching principles outlined in the preamble, emphasizing why recording is essential. Recording serves several crucial purposes, including:

Enhancing understanding: recording deepens our knowledge and perception of cultural heritage, offering valuable insight into its historical, architectural, and cultural significance.

Fostering participation: by engaging diverse audiences in the appreciation and preservation of cultural heritage, recording encourages broader participation from the public and private sectors.

Improve decision-making: accurate and comprehensive records support informed decision-making at all levels, ensuring that interventions are in harmony with the unique characteristics and qualities of heritage places.

Ensure respectful interventions: recording ensures that interventions are carefully planned and executed in a way that respects the designated characteristics and qualities of the cultural heritage monuments and sites.

Creating a robust record: recording establishes a permanent record of cultural heritage before any planned or unplanned changes occur, providing a valuable reference for future generations.

The process of cultural heritage recording and documentation is essential to the preservation of our collective history and identity. It plays a crucial role in the accurate understanding, protection and recognition of the immeasurable value of cultural heritage for present and future generations.

3.3.2 The Responsibility for the Recording of the Cultural Heritage

The second section of the ICOMOS document focuses on the responsibility for heritage recording. It examines the elements that influence or contribute to the responsibilities associated with the recording process, including commitment, character, and qualifications.

A commitment to preserving cultural heritage requires a corresponding commitment to recording. Those involved in the recording process should possess the appropriate qualifications, demonstrating the skills, expertise, knowledge and understanding required to ensure accurate recording. Training and lifelong learning programs are essential for achiving of these goals and objectives. Recognizing the interdisciplinary nature of recording, this document acknowledges the wide range of disciplines that may be involved in these activities. Responsibility also extends to managers, who must be credible and accountable for the use, development and preservation processes necessary to protect cultural heritage.

There is a gap exists in the lack of internationally recognized bodies authorized assess the adequacy of the qualifications for personnel involved in recording campaigns. It is essential that those conducting these tasks follow established recording principles, within a globally recognized framework. A notable example is the ICOMOS Principles for the Recording of Monuments, Groups of Buildings, and Sites. This internationally recognized document describes performance standards and expectations for recording activities.

This section of the ICOMOS document emphasizes that the responsibility for the recording of cultural heritage is shared among all stakeholders in the conservation process. Ensuring accurate and ethical documentation requires a high level of commitment and the consistent application of rigorous standards. This is essential for preserving the integrity of our cultural heritage, which reflects the diverse identities and histories of humanity.

3.3.3 Planning for the Recording of Cultural Heritage

The planning phase is one of the most crucial steps in the recording of cultural heritage. Planning involves deciding what to do and how to execute the recording campaign. In practice, this is a structured and systematic process that consists of carefully designed steps. A typical planning process could include steps similar to those illustrated in Fig. 3.3.

According to the ICOMOS Principles, there are two steps to follow when planning a recording campaign. The first step involves reviewing and evaluating existing records to determine their relevance and suitability for the specific recording campaign. The Principles provide recommendations on the types of documents

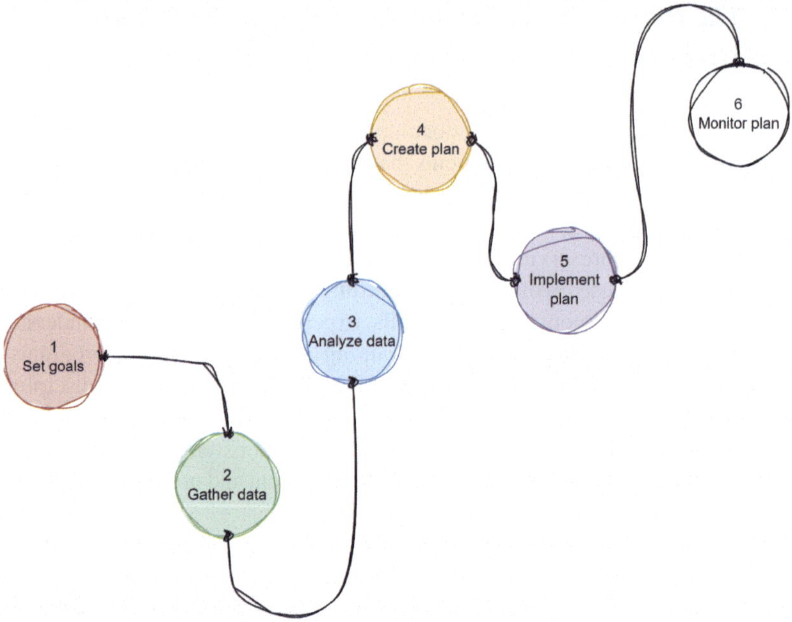

Fig. 3.3 A structured planning process

and resources that may be useful, emphasizing the importance of utilizing existing materials efficiently.

The second step involves defining the main criteria for selecting the appropriate scope, level, and methods of recording. Key criteria to consider when choosing suitable recording methods include the nature of the cultural heritage being recorded, the scope of the recording, and the available resources. Non-destructive techniques that preserve the integrity of the heritage site are preferred. Proper planning ensures that the recording process is comprehensive, accurate, and respectful of the cultural heritage's significance.

By following a structured planning process, practitioners can ensure that their recording campaigns effectively capture the necessary details while preserving the cultural heritage. Proper planning also allows for the efficient use of resources and facilitates the selection of the most suitable recording techniques for each unique case.

3.3.4 Content of the Records

Records are collections of data and information that are created, received, and maintained to provide evidence and verification of actions taken by individuals or organizations. The content of records is crucial for documenting the condition and structure of cultural heritage properties. This documentation is crucial not only for potential conservation measures, but also for capturing an accurate snapshot of the state of the property's state at a specific point in time.

A specific section is devoted to the content of records in the ICOMOS Principles. The document provides guidelines for the description of key identification parameters for cultural heritage properties, including location information, details of the recording process, and references to related records. The document also emphasizes the importance of documenting the sources of information used in the recording process. Additionally, a comprehensive checklist is provided to ensure the thoroughness of the recording.

The Principles also highlight the importance of matching the level of detail in the record to the needs of the intended users. This ensures that the information collected is appropriate for a variety of purposes, such as conservation, research, and public engagement. By adhering to these guidelines, practitioners will be able to create comprehensive and accurate records that effectively support the conservation and understanding of cultural heritage properties.

3.3.5 Management, Dissemination and Sharing of Records

Records management involves the efficient and systematic oversight of the creation, receipt, maintenance, use, and disposition of records. This includes processes for capturing and preserving of comprehensive facts and information about activities in the form of records.

The final part of the ICOMOS Principles focuses on the management, dissemination, and sharing of records. This section emphasizes the importance of proper management practices to protect records once they have been created. Proper management ensures that the records are preserved in their original form and remain accessible for use in the future. The Principles emphasize the importance of securing records to prevent loss, damage, or unauthorized access while ensuring they remain accessible to those with a legitimate interest.

This section also emphasizes the need to use appropriate technology to support effective management. Technological advances can greatly enhance the ability to efficiently store, manage, and share records. Properly managed records provide accurate and comprehensive information about cultural heritage, supporting conservation efforts, research, and public engagement. The Principles emphasize the need to balance security with accessibility, ensuring that records are both protected and available to those who need them.

3.4 The Convention for the Safeguarding of the Intangible Cultural Heritage

The *Convention for the Safeguarding of the Intangible Cultural Heritage* was adopted by UNESCO in 2003. It aims to protect and promote cultural heritage in its intangible forms. In contrast to tangible cultural heritage, such as buildings and monuments, intangible cultural heritage encompasses practices, representations, expressions, knowledge and skills that are passed down through generations. These cultural assets include oral traditions, performing arts, rites, festivals, and traditional crafts (UNESCO 2003). The Convention recognizes the importance of these practices for cultural identity, for cultural diversity, and for sustainable development.

At the heart of the Convention is the concept of community participation, which acknowledges that intangible heritage is a living expression, constantly evolving with its practitioners. This approach ensures that the measures taken to protect intangible heritage respect and enhance the cultural practices rather than imposing external standards. The Convention also established the Intangible Cultural Heritage Fund, which provides financial assistance to support efforts to conserve and protect intangible heritage, particularly in developing countries.

The Intergovernmental Committee for the Safeguarding of the Intangible Cultural Heritage meets annually, to evaluate nominations submitted by States Parties to the 2003 Convention. The Committee is empowered to decide whether to inscribe these intangible cultural practices and expressions on the Convention List. The purpose of establishing this list is to ensure better protection of important intangible cultural heritage worldwide and to raise awareness of its importance (UNESCO 2024a).

3.5 Treaties—Conventions—Principles

There are numerous international treaties, conventions, and principles that address ethics in professional practices. These documents serve as a framework for ethical conduct, guiding professionals to uphold standards consistent with ethical norms. They are crucial for several reasons:

Protection of human rights and promotion of social responsibility: by adhering to these principles, professionals and organizations can enhance societal well-being and ensure that their actions contribute positively to society. These guidelines help professionals navigate ethical dilemmas while ensuring that their practices uphold human rights and social responsibility.

Establishing a common set of ethical standards: these documents have gained international recognition as a framework for ethical conduct in professional practice. Adopting these standards ensures that professionals and organizations act in alignment with globally recognized ethical norms and values, promoting consistency in professional behavior.

Building trust and confidence: professionals and organizations committed to ethical standards help foster trust and confidence. Their adherence to ethical principles demonstrates a commitment to doing what is right, reinforcing their credibility and reputation among clients, colleagues, and the public.

Guiding decision-making: ethical standards provide invaluable guidance for decision-making, particularly in challenging situations. Adhering to established ethical principles ensures that decisions align with accepted norms and values, offering a clear path in complex ethical landscapes.

Together, these treaties, conventions, and principles provide a structured approach to ethical decision-making and ensure that professionals operate within an accepted ethical framework. This framework is essential for maintaining integrity, building trust, and promoting responsible professional conduct on an international scale.

3.5.1 ICOMOS Ethical Principles

ICOMOS was founded in 1965 in Warsaw, Poland, following the adoption of 13 resolutions by the Second Congress of Architects and Specialists of Historic Buildings in 1964. These resolutions aimed, among other objectives, to establish globally recognized principles for the ethical treatment of historic monuments. The first resolution, the Venice Charter, established the International Charter for the Conservation and Restoration of Monuments and Sites, and ICOMOS was created to implement this resolution. The Venice Charter drew its doctrinal roots from the 1931 Athens Charter on historic buildings, which provided a foundational definition of international cultural heritage and highlighted the importance of developing globally accepted ethical guidelines for the treatment of culturally significant sites.

As a global non-governmental organization, ICOMOS works to promote the conservation and protection of cultural heritage sites and monuments worldwide. In line with its mission, ICOMOS has developed a framework of *Ethical Principles* (ICOMOS 2020) to guide the responsible and ethical management and conservation of cultural heritage.

The foremost principle is to respect cultural heritage sites and monuments as invaluable expressions of human creativity and history. It is crucial to protect cultural heritage from damage and destruction while acknowledging the cultural, social, and economic values attached to it.

Minimal intervention to preserve the authentic material, form, and design of cultural heritage sites and monuments is a key concern. Conservation should take precedence over restoration, emphasizing the maintenance of original features and characteristics.

Furthermore, involving stakeholders, local communities, and experts in managing and conserving cultural heritage sites is essential. Community engagement activities and inclusive management approaches help ensure that the diverse values, needs, and perspectives of all stakeholders are considered in conservation efforts.

Transparency and accountability in the conservation and management of cultural heritage are crucial. This includes the implementation of clear and transparent decision-making processes and the dissemination of information about cultural heritage sites and monuments to the public and other stakeholders.

3.5.2 The APT Code of Ethics

The purpose of the Association for Preservation Technology (APT) is to "*advance appropriate traditional and new technologies to care for, protect, and promote the longevity of the built environment and to cultivate the exchange of knowledge throughout the international community.*" Membership in APT is open to a wide range of professionals, including preservationists, architects, engineers, conservators, consultants, contractors, craftspeople, curators, developers, educators, researchers, historians, landscape architects, students, and technicians. Any individual directly involved in applying methods and materials to preserve and protect historic structures and sites is eligible for membership.

APT members are committed to conducting their professional activities with honesty and integrity, ensuring that their work meets the highest ethical standards. They are also committed to respecting the values, history, and significance of cultural heritage in their work. Members are encouraged to collaborate and share information to promote the collective advancement of preservation technology knowledge. In addition, APT members are committed to maintaining and improving their professional competence through lifelong learning and knowledge sharing.

To fulfill APT's mission, its members adhere to a set of ethical commitments outlined in the APT Code of Ethics. This code provides a comprehensive framework to guide preservation technology professionals in maintaining ethical standards in their work, ensuring the protection and longevity of the built environment. It includes the following principles (APT 2019):

- *Pursue and promote the preventive care, examination, documentation, treatment and study of cultural property, guided by preservation philosophy, principles and skills.*
- *Demonstrate informed respect for the cultural property, its unique character and significance, and the people or person who created it.*
- *Consider, select, and utilize methods and materials to the best of current standards that will not adversely affect the integrity, function, investigation, or treatment of cultural property.*
- *Undertake only work for which they are qualified and prepared, engaging appropriately experienced specialists to conduct work for which they also are qualified.*
- *Provide advice, testimony and public statements only when informed on the specific circumstances of the matter and qualified to do so.*

- *Accurately represent their professional achievements and acknowledge others who contribute to work for which they are being recognized or seeking recognition.*
- *Seek to ensure the creation, maintenance, and dissemination of an appropriate record when preservation of significant cultural resources is not possible.*
- *Maintain their competence through continuing education and professional engagement to advance the field of preservation technology through the exchange of information and experience.*
- *Act with honesty and respect in all professional relationships, supporting the development, opportunities, acknowledgement, and fair treatment of colleagues.*
- *Minimize risks and hazards to co-workers, the public, and the environment.*
- *Use the resources of APT, as appropriate, for guidance in resolving ethical dilemmas and emergency situations.*

3.5.3 The AIC Code of Ethics and Guidelines for Practice

The American Institute for Conservation (AIC) is a professional association dedicated to conservators and allied professionals involved in preserving cultural heritage. The organization comprises thousands of members from various countries, all committed to promoting the preservation of cultural material. AIC envisions a world where cultural heritage is valued and supported, fostering a greater understanding and appreciation of its significance. The association advocates for the delivery of conservation services by qualified individuals who adhere to a strict code of ethics and practice guidelines.

According to (AIC 1994), the primary goal of conservation professionals is to preserve cultural property, which includes individual objects, structures, and collections. This cultural material may hold artistic, historical, scientific, religious, or social significance. Recognizing that cultural property is a priceless legacy that must be preserved for future generations, AIC members work diligently to protect it.

Conservation professionals have numerous responsibilities related to cultural property. The *AIC Code of Ethics and Guidelines for Practice* establish principles that guide all individuals involved in caring for cultural property. These principles provide a framework for maintaining high standards of practice, ensuring that conservation work is conducted ethically and in the best interests of the objects and collections being preserved (AIC 1994).

I *The conservation professional shall strive to attain the highest possible standards in all aspects of conservation, including, but not limited to, preventive conservation, examination, documentation, treatment, research, and education.*

II *All actions of the conservation professional must be governed by an informed respect for the cultural property, its unique character and significance, and the people or person who created it.*

III *While recognizing the right of society to make appropriate and respectful use of cultural property, the conservation professional shall serve as an advocate for the preservation of cultural property.*

IV *The conservation professional shall practice within the limits of personal competence and education as well as within the limits of the available facilities.*

V *While circumstances may limit the resources allocated to a particular situation, the quality of work that the conservation professional performs shall not be compromised.*

VI *The conservation professional must strive to select methods and materials that, to the best of current knowledge, do not adversely affect cultural property or its future examination, scientific investigation, treatment, or function.*

VII *The conservation professional shall document examination, scientific investigation, and treatment by creating permanent records and reports.*

VIII *The conservation professional shall recognize a responsibility for preventive conservation by endeavoring to limit damage or deterioration to cultural property, providing guidelines for continuing use and care, recommending appropriate environmental conditions for storage and exhibition, and encouraging proper procedures for handling, packing, and transport.*

IX *The conservation professional shall act with honesty and respect in all professional relationships, seek to ensure the rights and opportunities of all individuals in the profession, and recognize the specialized knowledge of others.*

X *The conservation professional shall contribute to the evolution and growth of the profession, a field of study that encompasses the liberal arts and the natural sciences. This contribution may be made by such means as continuing development of personal skills and knowledge, sharing of information and experience with colleagues, adding to the profession's written body of knowledge, and providing and promoting educational opportunities in the field.*

XI *The conservation professional shall promote an awareness and understanding of conservation through open communication with allied professionals and the public.*

XII *The conservation professional shall practice in a manner that minimizes personal risks and hazards to co-workers, the public, and the environment.*

XIII *Each conservation professional has an obligation to promote understanding of and adherence to this Code of Ethics.*

3.5.4 The Burra Charter: the Australia ICOMOS Charter for Places of Cultural Significance

The *Burra Charter* (ICOMOS 2013), first adopted in 1979 in the historic town of Burra, South Australia, has undergone several revisions, with minor changes in 1981 and 1988, followed by more substantial revisions in 1999. After a comprehensive review, the current version was adopted by Australia ICOMOS in 2013 (ICOMOS 2013).

The Burra Charter provides a framework of ethical and practical principles and guidelines for the conservation and management of culturally significant places (Fig. 3.4). It emphasizes the importance of understanding the cultural significance of a place and managing it in a way that respects that significance. The Charter advocates the use of appropriate conservation materials and techniques and highlights the importance of community involvement in the conservation process.

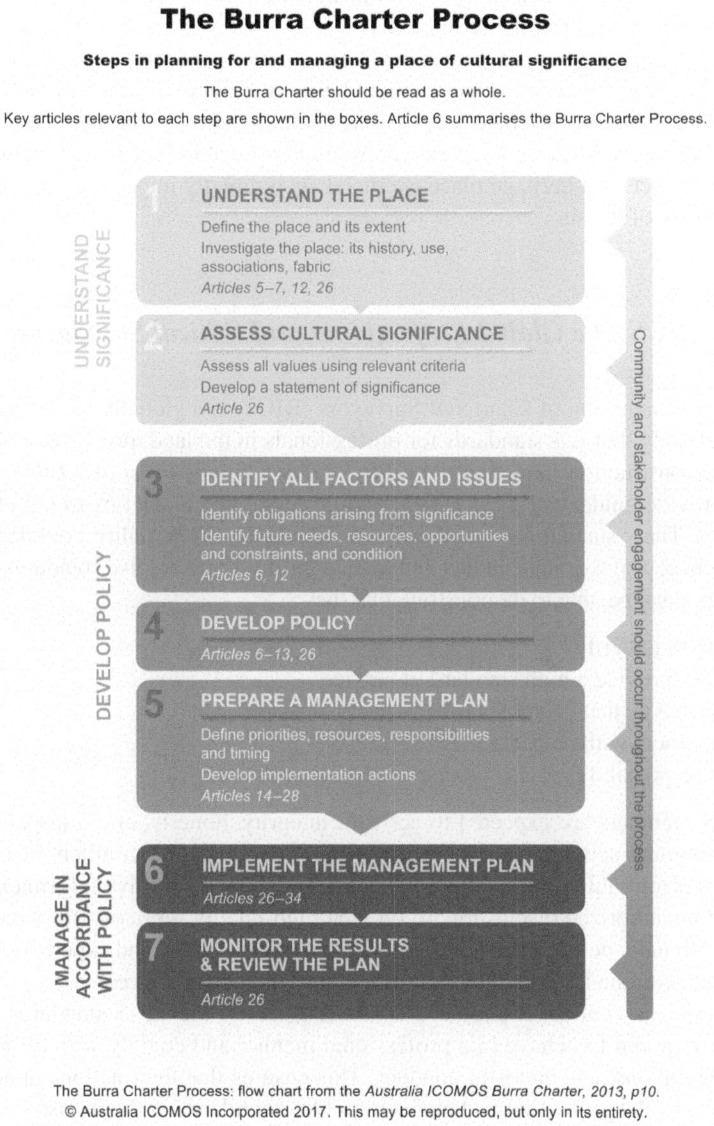

Fig. 3.4 The Burra Charter process: flowchart from the Australia ICOMOS Burra Charter, 2013

The Burra Charter has strong ethical dimension, recognizing that heritage places hold significant social and cultural meaning for people. Their conservation and management must be guided by ethical principles that are inclusive and consider the interests of all stakeholders, including the wider community.

A key ethical principle of the Burra Charter is the concept of *cultural significance*. In practice, this refers to the value that a cultural heritage holds for individuals or groups. The Charter acknowledges that cultural significance is not determined solely by physical attributes such as age or uniqueness but also by other factors, including cultural, social, and spiritual characteristics.

Another key ethical principle of the Burra Charter is *minimal intervention*. The Charter advocates that conservation should aim to preserve as much of the original material and character of a place as possible, while still allowing for necessary repairs and maintenance. Striking a balance between conservation needs and maintaining the significance of a heritage place is crucial, as it directly impacts the authenticity and integrity of the site.

3.5.5 RICS The Global Professional and Ethical Standards

The Royal Institution of Chartered Surveyors (RICS) is a globally recognized professional body that sets standards for professionals in the land, property, infrastructure and construction industries. The *RICS Global Professional and Ethical Standards* provide guidance on the ethical and professional conduct expected of RICS members. These standards cover areas such as integrity, impartiality, confidentiality, competence, professional conduct and sustainability. There are five standards and all members must be able to demonstrate that they:

1. Act with integrity
2. Always provide a high standard of service
3. Act in a way that promotes trust in the profession
4. Treat others with respect
5. Take responsibility.

RICS members are expected to act with integrity, honesty, and transparency in all professional activities, ensuring impartiality and avoiding conflicts of interest. They are also required to maintain the confidentiality of sensitive information and to avoid unauthorized disclosure. To provide high-quality services, RICS members must continually develop their professional knowledge, skills, and expertise, staying up to date with the latest industry developments and best practices.

Professional conduct is another important aspect of the RICS standards. Members are required to behave in a professional manner and comply with all relevant laws, regulations, and codes of conduct. This ensures that their actions uphold the reputation of the profession and maintain public confidence in the industry.

RICS also emphasizes the importance of sustainability in professional practice. Members are expected to consider the environmental, social and economic impacts of their activities and to promote sustainable practices that benefit both present and future generations. By adhering to these standards, RICS members contribute to a more ethical, responsible, and sustainable industry.

3.5.6 Seville Principles: International Principles of Virtual Archaeology

The *Seville Principles*, also known as the *International Principles of Virtual Archaeology* (Bendicho and Lopez-Menchero 2013), are a set of guidelines aimed at promoting the ethical and responsible use of digital technologies in archaeology. Developed during a meeting in Seville, Spain, in 2012, these principles are designed to guide the application of virtual archaeology tools and techniques, such as 3D scanning, photogrammetry, and virtual reality, in documenting, analyzing, and interpreting archaeological sites and artifacts.

The Seville Principles outline several key guidelines for the responsible use of virtual archaeology, including:

Transparency and openness of research: researchers are encouraged to be transparent and open about their methodologies, data, and processes. This ensures the reproducibility of results and fosters trust in the use of digital technologies in archaeology.

Respect for the archaeological record: it is essential to ensure that the application of digital technologies does not damage or misrepresent the archaeological record. The Principles advocate non-invasive methods that respect the integrity of archaeological sites and artifacts.

Collaboration between stakeholders: the Seville Principles emphasize the importance of collaboration between archaeologists, researchers, technology experts, and other stakeholders in virtual archaeology. This interdisciplinary approach ensures that virtual archaeological projects are scientifically robust and ethically sound.

Accuracy and reliability of data: virtual archaeology requires precise and accurate data to produce meaningful and reliable interpretations. The Principles highlight the necessity of applying rigorous standards in data collection and analysis to maintain the credibility of the field.

Preservation of sites and artifacts: the preservation of archaeological sites and artifacts should be supported by digital technologies. The Principles encourage the use of virtual archaeology to create digital records that contribute to the long-term preservation of cultural heritage for future generations.

Support for public education and outreach: virtual archaeology has tremendous potential for public engagement. The Principles advocate for the use of digital technologies to enhance public understanding of archaeology, making the field more accessible and engaging to a wider audience.

By adhering to these guidelines, researchers and professionals in virtual archaeology can ensure that their work is ethical, accurate, and respectful of the cultural heritage they aim to document and preserve.

3.5.7 *The London Charter for the Computer-Based Visualisation of Cultural Heritage*

The London Charter for the Computer-Based Visualization of Cultural Heritage, conceived in 2006, provides a set of guidelines designed to promote best practices in using computer-based visualization technologies in cultural heritage (Beacham et al. 2006). This charter resulted from a collaborative effort among experts from various fields, including archaeology, architecture, computer science, conservation, and cultural heritage management.

The Charter emphasizes the importance of the responsible and ethical use of computer-based visualization technologies in cultural heritage. It establishes a framework for the creation and dissemination of digital visualizations of cultural heritage sites and artifacts. This framework addresses several critical aspects of digital visualization technologies, including data acquisition, documentation, analysis, interpretation, and dissemination.

The website of the Charter (LondonCharter 2009) provides detailed information about these guidelines. To ensure that digital representations of cultural heritage maintain their integrity and respect the original contexts they are intended to represent, the Charter emphasizes transparency, accuracy, and sustainability in visualization projects. The Charter also advocates for the proper documentation of visualizations to allow for reproducibility and critical analysis.

By promoting these standards, the London Charter aims to build trust and confidence in digital visualizations by ensuring that they are produced in a scientifically accurate, ethically sound, and transparent manner. This is essential for the preservation of the cultural heritage that is represented by computational visualization technologies.

References

AIC. 1994. AIC Code of Ethics and Guidelines for Practice. American Institute for Conservation of Historic and Artistic Works. https://www.nps.gov/training/tel/Guides/HPS1022_AIC_Code_of_Ethics.pdf.

APT. 2019. APT Code of Ethics. APT College of Fellows/Standing Committee on Professional Recognition. Association for Preservation Technology. https://apt.memberclicks.net.

Beacham, R., H. Denard, and F. Niccolucci. 2006. An introduction to the London charter. In *The e-volution of information communication technology in cultural heritage—joint event CIPA/VAST/EG/EuroMed event*. Budapest: Archaeolingua.

Bendicho, V.M., Lopez-Menchero. 2013. International guidelines for virtual archaeology: The Seville principles. In *Good practice in archaeological diagnostics: Non-invasive survey of complex archaeological sites*, ed. C. Corsi, B. SlapšAak, and F. Vermeulen, 269–284. Springer. https://doi.org/10.1007/978-3-319-01784-6_16.

Charter, London. 2009. The London Charter: For the Computer-Based Visualization of Cultural Heritage. Version 2.1. http://www.londoncharter.org/. Accessed 15 Apr 2023.

ICOMOS. 1964. International Charter for the Conservation and Restoration of Monuments and Sites—Venice Charter. https://www.icomos.org/images/DOCUMENTS/Charters/Venice_Charter_EN_2023.pdf. Accessed 14 Nov 2020.

ICOMOS. 2012. Principles for the Recording of Monuments, Groups of Buildings and Sites (1996). https://www.icomos.org/en/charters-and-texts/179-articles-en-francais/ressources/charters-and-standards/387-principles-for-the-recording-of-monuments-groups-of-buildings-and-sites-1996. Accessed 14 Nov 2020.

ICOMOS. 2020. ICOMOS Ethical Principles. https://www.icomos.org/images/DOCUMENTS/Secretariat/Ethical_Principles/20150114-ethics-asadopted-languagecheck-finalcirc_rev202012.pdf. Accessed 03 May 2024.

ICOMOS, Australia. 2013. The Burra Charter: The Australia ICOMOS charter for places of cultural significance, 2013. Australia ICOMOS. https://australia.icomos.org/wp-content/uploads/The-Burra-Charter-2013-Adopted-31.10.2013.pdf.

UNESCO. 1972. The World Heritage Convention. http://whc.unesco.org/en/convention/. Accessed 14 Nov 2020.

UNESCO. 2003. Text of the Convention for the Safeguarding of the Intangible Cultural Heritage. https://ich.unesco.org/en/convention. Accessed 03 May 2024.

UNESCO. 2020a. List of World Heritage in Danger. http://whc.unesco.org/en/danger/. Accessed 14 Nov 2020.

UNESCO. 2020b. Tentative Lists. http://whc.unesco.org/en/tentativelists/. Accessed 14 Nov 2020.

UNESCO. 2020c. The World Heritage Committee. http://whc.unesco.org/en/committee/. Accessed 14 Nov 2020.

UNESCO. 2020d. World Heritage Fund. http://whc.unesco.org/en/world-heritage-fund/. Accessed 14 Nov 2020.

UNESCO. 2020e. World Heritage List. http://whc.unesco.org/en/list/. Accessed 14 Nov 2020.

UNESCO. 2024. Browse the Lists of Intangible Cultural Heritage and the Register of good safeguarding practices. https://ich.unesco.org/en/lists. Accessed 03 May 2024.

Chapter 4
Ethical Excellence

Abstract This chapter delves into the ethical frameworks and principles essential for the cultural heritage recording and documentation community. It outlines the necessity of adherence to international ethical standards and emphasizes the pursuit of ethical excellence in all aspects of cultural heritage preservation. The chapter explores various ethical challenges such as ownership, repatriation, data privacy, and commercialization of heritage sites, underscoring the importance of transparency, integrity, and community involvement. Aristotle's principles of virtue and ethical behavior provide a foundational perspective, advocating for balance, rational thinking, and the pursuit of happiness through ethical living. The chapter also examines the role of ethical principles in professional practice, highlighting fundamental values such as authenticity, confidentiality, fairness, and respect. Additionally, it discusses strategies for ethical decision-making and fostering an ethical culture within organizations, while addressing contemporary concerns including conflicts of interest and data security. Through structured approaches to ethical challenges and continuous improvement, professionals can achieve ethical excellence, ensuring trust and credibility in their work.

4.1 A Role Model for Ethical Excellence

The primary goal of this book is to provide the cultural heritage recording and documentation community with insights and a critical perspectives on the ethical considerations, data management, and preservation principles related to methods and techniques used in the field.

The book emphasizes two main objectives:

1. To ensure adherence to principles and standards established and endorsed by international organizations.
2. To strive for ethical excellence in all aspects of cultural heritage recording and documentation.

An ethical framework is essential for practitioners, researchers, end users, policymakers, businesses, and others interested in the field of cultural heritage recording

E. Stylianidis, *Exploring the Ethical Dimension in Recording and Documenting Cultural Heritage*, https://doi.org/10.1007/978-3-031-80034-4_4

and documentation. This framework supports the effective selection and application of existing solutions by integrating ethical considerations and fundamental legal perspectives. It encompasses the entire process from data collection and analysis to management and value enhancement.

With the rapid evolution of technology, users of digital solutions and data management within the field of cultural heritage recording and documentation frequently encounter significant challenges. These challenges include limited digital literacy, unfamiliarity with relevant legal frameworks, legal frameworks, and reluctance to adopt emerging technologies. Such obstacles can lead to gaps in information and communication, necessitating comprehensive solutions throughout the entire development cycle-from initial concept to the completed portfolio of captured and documented cultural heritage objects.

Applying the highest ethical, legal, and data management standards is essential in managing interactions with cultural heritage objects. Adherence to these standards is critical for several reasons:

- Protecting cultural heritage for future generations.
- Educating stakeholders involved in preservation processes.
- Ensuring the viability and sustainability of implemented solutions.
- Maintaining compliance with legal standards governing human and financial resources.
- Aligning with international, European and national regulations in a practical and sensible manner.

In addition to the various forms of human interactions (e.g., tactile, technological) and their associated ethical dimensions, there is also a need to anticipate the complex and interdisciplinary environments in which recording and documentation practices will be employed. This necessity arises from the involvement of multiple stakeholders, as well as from the organizational and budgetary challenges often associated with cultural heritage recording and documentation activities. Together, these considerations contribute to forming comprehensive preservation strategy.

Cultural heritage faces many ethical challenges, including issues of ownership and repatriation. Many artifacts are in stored far from their place of origin, resulting in complex legal and ethical issues regarding their rightful locations. The digitization of cultural heritage brings concerns about data privacy, access control, and potential misuse of digital replicas. Involving local communities in heritage projects ensures that their perspectives are adequately considered, which is particularly important when documenting sacred or sensitive sites. Commercialization of heritage sites presents ethical dilemmas by balancing between economic benefits with the need for preservation. Climate change and natural disasters pose significant threats to cultural heritage, making it imperative to ethically prioritize resource allocation for endangered sites.

Strict adherence to ethical guidelines is essential for successfully addressing these challenges. Transparency in the documentation process enables all stakeholders to understand the methods and decisions involved in preserving cultural heritage. Obtaining the informed consent of affected communities and respecting local customs are particularly crucial when dealing with sacred artifacts. Maintaining data integrity ensures the accuracy and authenticity of recorded information. This maintains the credibility of preservation efforts. Sustainable practices are essential to safeguarding cultural heritage sites and ensuring their accessible for future generations.

The ethical challenges of recording and documenting cultural heritage require a more robust and confrontational approach. This approach should not only comply with technical specifications and legal requirements established by national, European, or international authorities, and organizations but should also strive for ethical excellence.

Ethical excellence should not be viewed as merely *"nice to have"*, but rather as a *"must have"*, not only for human and social reasons but also to ensure the success of cultural heritage recording and documentation efforts aimed at preservation.

4.2 The Aristotelian Approach

In his treatise *Nicomachean Ethics* (Aristotle 2019), Aristotle (Fig. 4.1) discusses the pursuit of happiness through what he calls *'Ethical Excellence'*. This work is among the most influential ethical treatises in Western philosophy, wherein Aristotle examines various virtues and distinguishes between what is genuinely virtuous and what is mistakenly perceived as such. His assessment provides a framework that remains relevant in contemporary ethical discussions.

4.2.1 Ethics and Integrity

The importance of ethical behavior in personal and professional life, adherence to moral principles, and living by ethical standards are frequently debated topics. Despite the abundance of policies, ethical codes, and principles meant to guide behavior, adherence to ethics remains a significant challenge. As noted by (Ferrell et al. 2021), ethical violations occur in various settings, indicating a gap between understanding of ethical principles and consistently applying them.

Ethics involves studying principles that guide human behavior and distinguish right from wrong. This includes theories such as utilitarianism which focuses on the maximization of happiness, deontology which focuses on duty ethics, and virtue ethics which focuses on character qualities (Crane and Matten 2010). In addition, many professions have specific ethical guidelines that direct acceptable behavior within their fields.

Integrity refers to the quality of being honest and having a strong set of moral principles to uphold. According to (Kidder 2009), whether personal or professional, integrity involves consistently aligning one's actions with one's values. Individuals

Fig. 4.1 The statue of Aristotle adorns the ceremonial hall of the Aristotle University of Thessaloniki, Greece

who demonstrate integrity consistently act in accordance with their ethical beliefs, even in challenging situations or when unobserved.

Combined, ethics and integrity provide a framework for maintaining moral behavior in various aspects of life. Ethics provides the basic principles. Integrity ensures that these principles are upheld in practice. Both concepts are critical to the promotion of trust, fairness, and justice in society (Kidder 2009).

4.2.2 Understanding Ethics

Historically, the term ethics derives from the Greek word *ethos*, meaning the customs, habits, and norms of people or society as a whole. According to (MacIntyre 2016), understanding the true meaning of the word ethics can be difficult to grasp. It may be helpful to first understand what ethics is not before examining what it is. Ethics is not merely the observance of a set of rules or regulations; it is a deeper commitment to the values that form the foundation of human behavior.

Ethics is a philosophical discipline that examines the moral principles governing human behavior. It asks fundamental questions such as what is right and what is wrong, and what human beings should do in different situations. The study of ethics includes normative ethics, which determines what people should do, metaethics, which examines the nature of moral judgments, and applied ethics, which addresses controversial issues such as abortion and the death penalty. Through this lens, ethics provides guidelines that can help individuals navigate the moral complexities of their daily lives.

There are different frameworks for determining the right course of action within normative ethics. Utilitarianism, for example, holds that the best action is the one that maximizes overall happiness or minimizes suffering (Bentham 2007). Deontology, on the other hand, emphasizes the observance of duties and rules, regardless of the consequences. Virtue ethics focuses on the character and virtues of individuals rather than on specific actions. It promotes qualities such as honesty and kindness.

Ethics is critical in many professions as it provides guidelines to ensure responsible and fair behavior. Professional ethics are an integral part of maintaining public trust and ensuring that practitioners are accountable for their actions.

Rooted in centuries of philosophical thought, ethics serves as a compass for human behavior. By offering different frameworks to navigate the diversity of human situations, it seeks to answer complex moral questions. Through the understanding and application of ethical principles, individuals and societies can strive for actions and policies that promote justice, fairness, and the well-being of all.

4.2.3 Aristotle's View on Ethical Behavior

Aristotle attributes ethos to human activities. Ethos is often associated with negative connotations. According to (Shields 2004), he advises the avoidance of behaviors associated with wickedness, brutality, and lack of self-control. Instead, Aristotle advocates the orientation of one's actions toward virtue and restraint. The essence of ethical behavior, according to Aristotle, is striking a balance between habitual behavior and the guiding principles of action. It is this balance that fosters genuine desires and rational thinking in the soul. (Pakaluk 2005) notes that, according to Aristotle, moral virtue is the most practical path to effective action. When someone of virtuous character desires something and considers it as an ultimate goal through rational thinking, they experience it as beautiful. In this way, virtuous people have a keen sense of perception and sound judgment, because the true nature of beauty is visible only to those who have a good character.

The concept of the mean, or golden mean, lies at the heart of the Aristotelian view of ethical conduct. For Aristotle, ethical behavior isn't just about following a set of rules; it's about developing the right character through habit. He believes that by continually practicing virtuous actions, individuals cultivate a virtuous character that naturally leads them to make moral decisions (Shields 2004). This is consistent with his belief that ethical behavior is fundamentally about finding a balance between lack and excess. Courage, for example, is the virtue that lies between the extremes of being reckless (an excess) and being cowardly (a deficiency).

Aristotle emphasizes that rational thought is a crucial aspect of ethical conduct and behavior. He argues that practical wisdom (*phronesis*) is the guiding principle and that the moral virtues are inextricably linked to the intellectual virtues. According to (Pakaluk 2005), practical wisdom is the ability of an individual to discern the right course of action in a particular situation, considering the specific circumstances and the principles of virtue. Ethical behavior, then, is the alignment of one's desires and actions with rational reasoning, leading to actions that are in harmony with the individual's virtues. This harmony is what Aristotle describes as the achievement of *eudaimonia*, or flourishing. Eudaimonia is the highest form of human fulfillment.

4.2.4 The Relationship Between Virtue and Happiness

Aristotle argues that happiness, or eudaimonia, is the ultimate goal of human life. The attainment of this state is the result of a life lived in accordance with the principles of virtue. For Aristotle, virtue is not just a moral quality but an excellence that enables individuals to achieve their potential. Ethical excellence requires the alignment of desires with reason, and this harmony leads to a life of fulfillment.

He makes a distinction between the moral virtues, which govern actions and emotions, and the intellectual virtues, which guide thinking. Moral virtues require habituation: one becomes virtuous by performing virtuous acts repeatedly until they become second nature. The development of intellectual virtues, on the other hand, is a matter of education and time (Ackrill 1981).

In Aristotle's view, virtues provide a path to happiness by enabling individuals to live in accordance with their true nature as rational beings. The proper function of human beings is to live according to reason, and virtue helps to align what we want with rational thought. According to (Pakaluk 2005), this alignment allows individuals to bring their lives into harmony with the world around them, and to act with purpose and deliberation.

Eudaimonia, then, is not a fleeting pleasure, but rather a sustained state of well-being that results from living a virtuous life. It is the culmination of a life lived according to reason and virtue, where individuals reach their highest potential and find satisfaction in the meaningful pursuit of the good. This concept emphasizes the intrinsic link between virtue and happiness in Aristotle's philosophy: virtue is not merely a means to happiness, but is a core aspect of it, emphasizing that the virtuous life is itself a rewarding and fulfilling end.

4.2.5 *Ethical Excellence in Modern Contexts*

Aristotle's ideas continue to remain highly relevant in discussions of contemporary ethics. In today's fast-paced world, the need for ethical excellence in a variety of fields such as business, politics, and technology is becoming increasingly apparent. Ethical failures, often driven by short-term gain or pressure, reveal the need for deeply rooted virtues that bring intentions and actions into alignment. Aristotle's focus on habituation—the practice of virtuous actions until they become second nature—guides.

A practical application of Aristotelian ethics today is in the area of leadership. According to (Solomon 1993) ethical leadership involves using virtues such as honesty, courage, and justice to guide decisions and actions. It goes beyond compliance with laws or rules. It emphasizes the importance of character in building trust and fostering sustainable practices. Leaders who cultivate and demonstrate these virtues create an environment in which ethical behavior is valued and has an impact not only on their organizations but also on society as a whole.

In the realm of technology, Aristotle's ethical framework can be used to address emerging challenges related to data privacy, artificial intelligence, and digital ethics. The virtuous use of technology requires not only technical competence but also the exercise of moral discernment. (Vallor 2016) suggests that this involves evaluating the ethical implications of new technologies and ensuring that they are consistent with human flourishing. Through the cultivation of intellectual virtues such as wisdom and moral virtues such as justice, individuals can navigate these complex issues in a thoughtful manner.

In addition, in the realm of business, ethical excellence is critical to the maintenance of trust and reputation. Aristotle's emphasis on practical wisdom (phronesis) and the importance of aligning actions with virtuous intentions can guide companies in making decisions that consider the broader impact on society. As (MacIntyre 2016) argues, ethical companies that consistently apply these principles can foster loyalty among their stakeholders and make a positive contribution to the communities in which they operate.

4.3 Ethics in Professional Practice

Ethical principles are incredibly valuable, not only in the personal lives of individuals but also in their professional lives. These principles form the basic beliefs and standards that facilitate loyalty and smooth operation in the professional environment and shape the way we interact with each other in society. Without ethical norms, society would experience widespread dishonesty and insecurity. This would create a challenging situation that would lead to conflict and a lack of harmony. For example, some professionals might deliberately withhold or misuse information for personal gain rather than for the benefit of cultural heritage if ethics didn't apply to the recording and documentation of cultural heritage.

Engaging in unethical behavior clouds the mind and distorts the results of professional work. Truth is distorted and reality is hidden from society's view when we engage in unethical behavior in our daily professional activities. The minds of unethical professionals become confused and unclear, with consequences not only for their self-image but also for their reputation in the eyes of others. Ethical principles in our daily activities and professional operations serve as a guide for achieving honorable and credible results and developing social norms that ultimately affect our collective future, especially for future generations.

In the field of ethics, there are two main issues on which the concern is primarily focused:

1. What kinds of actions should we take?
2. What kinds of actions should we avoid?

These are questions of vital importance which must have answers. Before we can address them, however, we need to clarify what we mean by "good" and "bad" practices. This will allow us to determine how ethical behavior in professional work relates to the pursuit of good practices and the avoidance of bad practices.

The code of professional ethics serves as the foundation upon which any reputable profession is built. It comprises a comprehensive framework of moral and ethical principles. It is carefully designed to guide the conduct and decision-making processes of professionals in their respective fields. At its core, ethics in professional life serves a dual purpose: to ensure that professionals consistently adhere to certain standards of conduct, while at the same time fostering trust and maintaining the integrity that is the foundation of their vocation.

Professionals use specialized knowledge and skills when providing a service to an individual, the public, or the private sector. Specialized training acts as a catalyst in decision-making. Because of the specialized training they have received at different levels of education (e.g., undergraduate or postgraduate), professionals are skilled at making judgments and decisions. In this way, they are able to practice their skills and make informed decisions under conditions that are generally not easily accessible to the public at large.

Codes of conduct have been put in place by many professions. This internal process for the enforcement of codes of conduct is followed by a large number of professional organizations. The purpose is to encourage members of the profession to follow the code of conduct, to prevent exploitation of the customer and to protect the integrity of the profession. Professional ethics encompasses many types of standards, including those related to individual, organizational, and corporate standards of how professionals practice their profession according to the accepted principles that serve as a rule for a profession.

The Hippocratic Oath is an ethical oath taken by physicians. It is considered to be one of the earliest examples of a professional code of ethics. Written by Hippocrates, a Greek physician considered one of the most influential figures in the history of medicine, the oath is still held sacred by physicians. The oath addresses issues such as treating the sick to the best of one's abilities, protecting patient privacy, and teaching and passing on medical knowledge to the next generation.

These types of standards not only benefit the individual or organization that uses the services but also give the individual a real sense of belonging to the profession. Standards at the disciplinary level allow the profession to establish a standard of conduct and to make sure that individual professionals are held to that standard. This attitude enables conscientious professionals to practice knowing that they will not be undercut by less conscientious professionals. It also maintains public confidence in the profession. This encourages the public to continue to seek the services of the profession.

4.3.1 Fundamental Ethical Principles in Professional Practice

Ethical principles serve as a timeless compass to guide professionals through the complexities of modern life in a rapidly evolving world where technology and globalization are reshaping industries and societies alike. According to (Trevino and Nelson 2021), despite the unique challenges faced in different fields, professionals adhere to a common set of ethical standards. These standards maintain the integrity and sustainability of their professions. These principles are the foundation for trust, collaboration, and excellence within and across industries, as well as shaping individual behavior.

While ethical principles are often simple in concept, they can be challenging to uphold amid the pressures of day-to-day professional life. Individuals are often tempted to compromise their values due to financial incentives, competitive environments, and evolving regulations (Crane and Matten 2010). Adherence to ethical standards, however, is critical to maintaining credibility and trust. Without them, the fabric of professional integrity begins to fray. This leads to mistrust and diminishes the quality of services provided.

The importance of these principles is further underscored by the interconnectedness of today's global economy. A breach of ethics in one profession can have ripple effects across industries and affect the lives of countless people. Failing to provide transparency in financial institutions can lead to economic crises while failing to maintain confidentiality in healthcare can damage patient trust and safety (Ferrell et al. 2021). As a result, ethical behavior is not just a personal or organizational concern; it is a societal imperative.

The importance of ethics has also increased in the digital age. The ease with which information is disseminated requires greater vigilance in protecting privacy and confidentiality. The rise of artificial intelligence and automation in the Industry 4.0 era further complicates ethical decision-making. Professionals must continually refine their understanding of right and wrong in uncharted territory.

Understanding and practicing core ethical principles becomes paramount. This chapter explores these core principles and how they guide professionals to maintain integrity, foster trust, and promote public good. All professionals, from seasoned practitioners to emerging leaders, can benefit from a renewed focus on these core ethical values.

Authenticity in a professional context is a matter of being true to oneself while at the same time being true to ethical principles. It encompasses honesty in actions, communication, and intentions. Authentic professionals are transparent about their abilities and limitations, fostering genuine relationships and building reputations.

Confidentiality is a commitment to protecting sensitive information obtained through professional activities. Whether it's customer information, business strategies, or personal details, professionals respect the privacy of others by not disclosing this information without proper authorization. This principle is critical to maintaining trust and confidence and protecting individuals from harm.

Connectedness stresses the importance of relationships in the professional world. Professionals recognize that they are interdependent with colleagues, clients, and the community at large. Through cultivating positive relationships and collaboration, they contribute to a supportive environment that benefits all stakeholders.

Credibility is the foundation upon which trust and respect are established. It is earned through consistent ethical behavior and a proven track record of being reliable and trustworthy. Professionals enhance their credibility by consistently delivering on their promises, adhering to standards, and demonstrating their expertise in the field they serve.

Fairness means treating all people in a fair and impartial manner, without discrimination or favoritism. While recognizing diversity and being sensitive to different needs and backgrounds, professionals apply the same standards and rules to everyone. This ethical attitude helps prevent conflict and promote justice.

Gratitude is a powerful principle in promoting positive relationships, but it is often overlooked in professional ethics. A culture of kindness and generosity is fostered by expressing appreciation for others' contributions and acknowledging support. The practice of gratitude helps professionals maintain a positive outlook and build stronger relationships.

Honesty is closely related to integrity; honesty includes the provision of truthful, accurate, and complete information in all professional dealings. An honest professional discloses pertinent facts and avoids misrepresentation or omission of information. This transparency ensures that others can make informed decisions. It also contributes to a fair and equitable environment.

Integrity which reflects a commitment to act honestly and consistently with moral principles, is the cornerstone of professional ethics. By upholding the highest standards of conduct in every situation, professionals demonstrate integrity by matching their actions with their words. This fosters a culture of honesty, and accountability that builds trust with colleagues, clients and the community.

Lifelong learning complements professional competence through the recognition of the need for continuous education. The rapid evolution of practices, technologies, and tools, makes it necessary for professionals to update their skills and knowledge on an ongoing basis. This mindset fosters innovation, adaptability, and sustained professional growth.

Objectivity requires that professionals remain unbiased in the judgments and decisions they make. Rather than being influenced by personal feelings, interests, or external pressures, they must rely on facts, evidence, and rational reasoning. By adhering to objectivity, professionals ensure that all those involved receive fair treatment and maintain the integrity of their work.

Privacy goes beyond confidentiality and emphasizes the right of individuals to have control over their personal information. Professionals respect this right by collecting, using, and disclosing information only for legitimate purposes and with the consent of the individual. This ensures that personal freedoms and autonomy are respected in the professional context.

Professional competence involves maintaining the knowledge and skills necessary to perform one's job effectively. This requires staying abreast of the latest developments in one's field of expertise, understanding the relevant standards, and consistently applying best practices. Accomplished professionals deliver high-quality services and inspire confidence in in their abilities.

Respect is the recognition of the inherent dignity and worth of every individual. Professionals who practice respect demonstrate consideration for the ideas, beliefs, and boundaries of others, even when differences of opinion arise. This principle promotes inclusiveness, open-mindedness, and an appreciation of diversity, which are essential for creating a professional environment that is harmonious and productive. Respectful professionals listen actively, value what others contribute, and foster an atmosphere where all feel valued and understood.

Transparency means being open and clear about how things work, about decisions made, and about intentions. By providing full disclosure and engaging in open communication, professionals ensure that stakeholders understand the rationale behind the actions they take. This clarity forms the basis for trust and helps prevent misunderstandings and conflicts.

Trust is the culmination of all of these principles and represents the confidence that others have in the reliability, integrity, and competence of a professional. Building trust requires consistent ethical behavior over an extended period of time. Professionals create a trustworthy environment that fosters collaboration, satisfaction and success when they embrace these principles.

These fifteen fundamental ethical principles (Fig. 4.2) serve as a guiding beacon for professionals in a wide variety of disciplines. By committing themselves to these values, individuals will not only enhance their own reputations, but will also contribute to the integrity and vitality of their professions and of society as a whole.

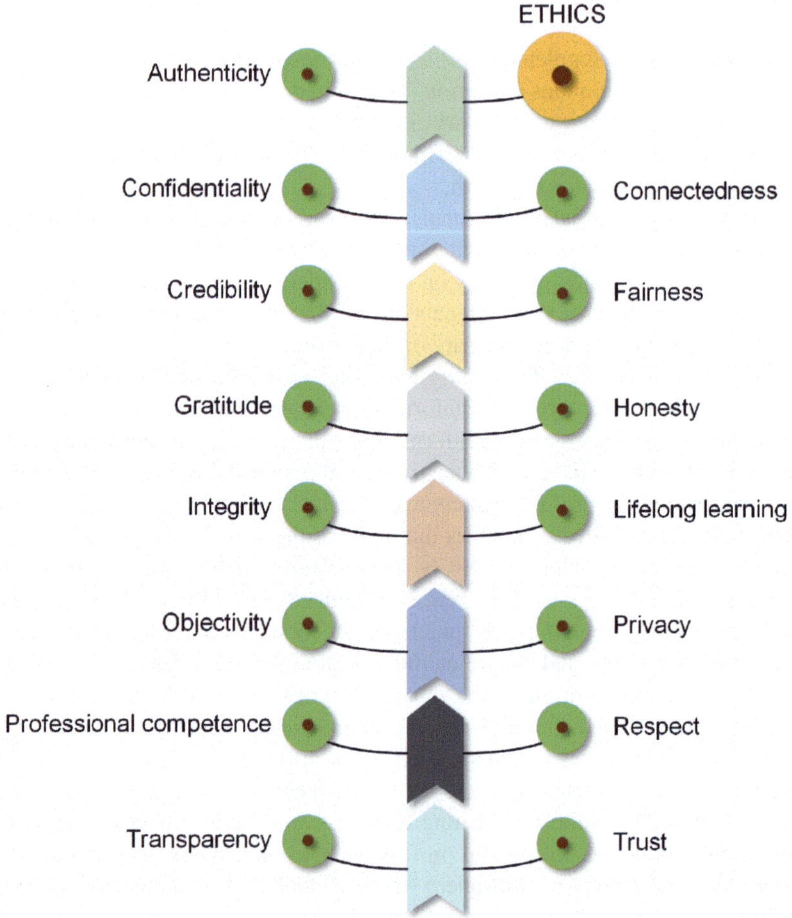

Fig. 4.2 Fundamental ethical principles in professional practice

4.3.2 *Ethical Challenges and Dilemmas in Professional Practice*

In today's interconnected world, professionals in a variety of disciplines face a myriad of ethical challenges stemming from advances in technology, globalization, and evolving societal norms. The pace of change in business, healthcare, technology, and other fields has created new and complex ethical landscapes. These landscapes require professionals to continually adapt. Traditional ethical guidelines may no longer be sufficient in this rapidly changing environment, necessitating the development and implementation of updated frameworks that address today's challenges. As a result, a strong ethical foundation is critical to maintaining trust and confidence, upholding professional standards, and ensuring that responsible decisions are made.

Conflicts of Interest

Managing conflicts of interest, which often arise from overlapping personal and professional commitments, is one of the key challenges of modern professional practice. It is imperative that professionals recognize and manage these conflicts, as the increasing interconnectedness of global markets and organizations can blur the lines between personal gain and professional duty (Boatright et al. 2017). In these situations, professionals need a strong ethical foundation to manage dilemmas effectively, even under intense pressure and scrutiny.

Conflicts of interest arise when professionals face situations in which their personal interests may interfere with their impartiality or judgment. To manage these situations, transparency and strict adherence to ethical guidelines are essential to prevent conflicts from interfering with the best interests of the client.

To effectively manage conflicts of interest, professionals should prioritize transparency. They should disclose any affiliations that could influence their recommendations. Providing of comprehensive information to clients is also a prerequisite for enabling them to make informed decisions. In some cases, an independent audit or review may be necessary to ensure objectivity and maintain trust. This can help professionals navigate complex situations with integrity (Ferrell et al. 2021).

Data Privacy and Security

In the digital age, where confidentiality remains a cornerstone of professional ethics, data privacy and security have become issues of paramount importance. As technology continues to advance at a rapid pace, maintaining privacy has become increasingly challenging, as professionals often need to share client information for legitimate purposes without compromising its security. The proliferation of electronic data stored and transmitted has created vulnerabilities and requires professionals to be vigilant regarding data breaches (Solove 2010). As the value of sensitive information increases, the importance of protecting that information also increases, requiring robust privacy measures and a strong ethical obligation.

In the real world, ethical dilemmas often arise when trying to balance the need to share information with the need to protect individual privacy. Reporting the breach could have serious consequences for the employee, impacting their career and reputation. This scenario illustrates the delicate balance that professionals must strike between protecting individual privacy and maintaining ethical standards within their organizations.

To deal with such dilemmas, it is important to take a structured approach to resolving them. The first step is to assess the severity of the data breach and understand its potential impact (Schwartz and Solove 2011). This assessment will serve as the basis for determining the urgency of the situation and the appropriate course of action. The second step is to discuss the issue with the affected colleague to understand the context and to ensure that any subsequent actions are based on a clear understanding of the situation. Finally, ensure that corrective action is both effective and ethical by following established protocols to address the violation while maintaining confidentiality. By following these strategies, professionals can navigate the complex landscape of data privacy and security. They can protect sensitive information and maintain ethical standards.

Professional Responsibility Versus Client Loyalty

Navigating the delicate balance between professional responsibility and loyalty to a client is a significant ethical challenge for many members of the profession. Lawyers, for example, are often face scenarios in which their duty to maintain the confidentiality of their clients may conflict with their legal or societal duty to report illegal activity. These conflicting responsibilities can place them in a position where they must carefully weigh their loyalty to their clients against their broader obligations to the legal system and society as a whole. Achieving of this balance requires a nuanced understanding of the legal and ethical framework within which the profession operates (Rhode 2013).

When faced with such dilemmas, professionals should first conduct a thorough review of the legal obligations and ethical guidelines applicable to their particular situation. These frameworks provide a foundation for navigating complex decisions and highlight the consequences of different courses of action. Seeking advice from ethics committees or legal counsel is essential in situations where the right course of action is not immediately clear. Such guidance can help professionals clarify their responsibilities and determine the most ethical way to proceed, particularly in scenarios where personal and professional interests may intersect.

Ultimately, professionals need to consider the broader implications of what they are doing, not only for the client but also for the profession and society at large. This holistic perspective will ensure that decisions are made in accordance with ethical standards and reflect a commitment to upholding the integrity of the profession. By following these strategies, professionals can effectively navigate the challenging landscape of balancing responsibility and loyalty and ensure that they are acting in an ethical manner while at the same time serving the best interests of their clients and the public.

4.3.3 Ethical Decision-Making in Professional Practice

Ethical decision-making in professional practice requires a structured approach to deal with the complex dilemmas encountered by professionals. Recognizing and defining the ethical issue, which often involves understanding the context and potential impact of the decision on various stakeholders, is one of the first steps in this process. Proper framing of the ethical dilemma allows for a clearer understanding of the challenges and constraints, which is essential to the evaluation of possible solutions. This step is followed by the evaluation of the stakeholders involved, the consideration of alternatives, and the making of a reasoned decision. Reflecting on the results is essential for continuous learning and improvement once the decision has been made and implemented.

The model developed by (Rest 1986), which provides a systematic framework for resolving ethical dilemmas, is an essential tool for ethical decision-making. This model includes the recognition of the moral issue, the formation of a moral judgment, the determination of moral intention, and the performance of moral behavior. By following this model, professionals are able to systematically address the nuances of an ethical dilemma while ensuring that no important considerations are overlooked (Rest 1986). This method helps professionals distinguish between competing values, to prioritize ethical principles, and reflect on the implications of the decisions they make.

It is important to consider the potential consequences of each option, not only for the immediate parties involved but also for society as a whole when making ethical decisions. This requires professionals to evaluate their decisions against professional standards, social norms, and relevant legal requirements. Furthermore, consultation with colleagues or ethics committees can provide additional perspectives and insights that can be helpful in challenging any biases that may influence the decision (Ferrell et al. 2021). This type of consultation can also help to strengthen the integrity of the decision-making process.

In practice, ethical decision-making is an iterative process that involves continuous reflection and adjustment. In order to understand the real-world implications of their choices and to refine their ethical judgments for the future, professionals must reflect on the consequences of their decisions after they have been implemented. This reflection is essential to learning from both successes and failures, which will enable professionals to deal more effectively with ethical dilemmas in the future (Ferrell et al. 2021). Thus, the adoption of a structured approach to ethical decision-making can help to ensure that professionals act in a manner consistent with their values and uphold the standards of their profession.

4.3.4 Developing an Ethical Culture in Professional Practice

Leadership has a critical role to play in setting the tone for an ethical culture within an organization. Leaders serve as role models. Their actions reflect the ethical values of the organization. Their commitment to ethical principles has a profound effect on the culture of the organization, which often cascades down through all levels of management. (Ferrell et al. 2021) emphasize that effective leaders create an environment where ethics are openly discussed, prioritize ethical behavior, and instill it into the organizational ethos. This can be achieved through the consistent promotion of ethical policies, by holding individuals accountable for their actions, and by maintaining transparency in the decision-making process. Encouraging an open dialogue about ethics ensures that employees feel comfortable raising ethical concerns and know that their perspectives are valued and will be addressed.

A clear and comprehensive code of ethics serves as a foundational document that provides a road map for the behavior that is expected within an organization. It articulates the values, principles, and ethical standards that the organization upholds and provides guidance to professionals as they navigate challenging situations. The development of a code of ethics requires the involvement of stakeholders throughout the organization to ensure that the code resonates with the culture and challenges that are specific to the organization. Implementing this code goes beyond publishing a document and means embedding the code into the way the company operates, from the induction of new employees to the decision-making process. Reviewing and updating the code on a regular basis ensures that it remains relevant and effective as a guide to ethical behavior.

Training is essential to help professionals recognize ethical dilemmas and respond to them in an effective manner. Regular ethics training ensures that employees are kept abreast of evolving standards and are able to confidently navigate the ethical challenges that arise in their day-to-day work. Training programs should be interactive and scenario-based, allowing participants to practice making decisions in the context of simulated ethical dilemmas (Ferrell et al. 2021). This fosters a deeper understanding of ethical expectations and helps reinforce the principles outlined in the code of ethics. By ensuring that individuals are prepared to make informed and principled decisions, continuous development of ethical competence keeps ethics at the forefront of professional practice.

The promotion of ethical behavior within teams and organizations is critical to the establishment of an ethical culture. Recognize and rewarding ethical behavior, reinforces its importance and motivates others to follow. Additionally, the establishment of systems for the safe reporting of unethical behavior without fear of retribution promotes transparency and accountability. Such systems should provide clear guidelines on how to report concerns and ensure confidentiality to protect those coming forward. In order to effectively and impartially address reports of unethical behavior, organizations should also implement a comprehensive response mechanism. Through the promotion of ethical behavior via these mechanisms, organizations can foster a culture in which integrity is upheld and ethical behavior is the norm.

4.3.5 Regulatory and Legal Aspects of Ethics in Professional Practice

Ethical practice in professional settings is guided by a variety of legal and regulatory frameworks that vary from industry to industry to address the unique challenges and requirements of each. According to (Hartman et al. 2024), regulatory bodies play an integral role in creating and enforcing ethical standards tailored to the specific needs of different professions. Some organizations promote consistency and accountability while protecting the interests of clients, and the public by developing codes of conduct that outline the expected ethical behavior of professionals.

Multiple enforcement mechanisms, such as licensing, certification, and disciplinary action, are used to ensure compliance with these ethical standards. According to (Shaw 2016), regulatory bodies conduct audits and reviews on a regular basis to ensure that professionals are adhering to the established ethical framework. They also impose disciplinary sanctions for violations. These can range from public reprimands to suspension or revocation of licenses. To maintain ethical standards and ensure compliance, such enforcement mechanisms serve as both a deterrent and a corrective tool. This oversight is essential to the maintenance of ethical conduct and the preservation of the credibility of the professions that these organizations oversee.

Professionals who violate ethical standards can face serious consequences. These consequences can have a significant impact on their careers and reputations. Consequences can range from fines and sanctions to loss of professional licenses. Loss of professional licenses can prevent individuals from practicing in their field. In addition, because trust and confidence are so important in professional interactions, a damaged reputation may result in long-term professional consequences (Jennings 2017). The importance of ethical behavior and integrating ethical principles into everyday practice is highlighted by the potential consequences of misconduct. This ensures that professionals act in the best interests of their customers, their institutions, and the broader community.

4.3.6 Future Trends in Ethics in Professional Practice

Emerging technologies, in particular artificial intelligence (AI) and big data, are presenting a new set of ethical challenges in the practice of the profession. The vast amounts of data that are collected require strict measures to protect privacy, as unauthorized access to or misuse of data can have significant consequences (Floridi et al. 2018). Furthermore, biases embedded in algorithms can lead to unfair outcomes, and AI systems raise concerns about decision-making. As a result, professionals in tech-driven fields will need to establish frameworks for transparency, accountability, and data governance to ensure ethical behavior (Mittelstadt et al. 2016).

With globalization, professionals often find themselves working in environments where the ethical norms are very different. What is considered acceptable in one culture may be considered inappropriate in another. This requires a diversified understanding of cultural differences. Professionals must have a commitment to continuous learning in order to appreciate different values and to navigate these complexities thoughtfully. A universal code of ethics may provide some guidance, but respect for local customs and ethical principles remains of paramount importance.

Significant challenges to ethical practice that require constant vigilance and adaptability arise from the rapid pace of technological advancement. Professionals need to keep abreast of emerging trends in environmental sustainability and social justice to incorporate these considerations into their ethical frameworks. In addition, the ethical challenges posed by rapidly evolving technologies require innovative approaches to anticipate and address potential problems before they manifest themselves. This presents an opportunity for professionals to lead by example in developing responsible solutions.

Ethical practice in professional contexts is being reshaped by the emergence of new technologies, globalization, and changes in society. Professionals are faced with the challenge of maintaining ethical integrity in an environment of evolving norms and rapid advances in technology. However, these challenges also provide opportunities for leadership in ethical innovation. As long as professionals remain vigilant in their adherence to ethical principles and are willing to adapt to changing landscapes, they can lead the way.

4.4 Ethical Excellence—Some Concluding Thoughts

The relationship between employees and organizations is analogous to that of a couple. Employees are expected to respect their colleagues and act in a professional manner that is both effective and productive. However, this does not imply that they should compromise their ethical values. Excellence in work is achieved when employees act with good intentions and in the interest of the greater good. This not only benefits the organization but also sets an example for others. Ethical excellence is a sustainable and fulfilling outcome.

Excellence is the result of individuals who are mobilized by a visionary workflow to generate constructive, productive, and valuable outcomes. These outcomes contribute to effectiveness and fulfillment in task execution. This condition demands a high level of readiness, especially on an interpersonal level. In practice, human fulfillment arises from performing the best one can, necessitating the maximization of charisma, skills, talents, and individual competence to carry out tasks at the highest potential.

Individuals should strive to operate at their peak potential, as excellence is an unstoppable and constant endeavor to achieve this goal. Excellence implies the status or quality of being outstandingly good or immensely notable. It denotes a perpetual and vigorous attempt to advance the boundaries of accomplishment as well as fulfillment. "*Excellence is not an act, but a habit*," is attributed to Aristotle. The habit he referred to is that of tangling the whole being in activities.

Adherence to moral values and ethical principles during the execution of tasks leads to ethical excellence. This is defined as the achievement of efficacy and fulfillment during the performance of professional-level duties. Stakeholders require a stable environment of trust to build upon, and trust can be delivered through ethically excellent work behavior based on authenticity, credibility, quality, and transparency.

The core ethical values are considered the original source for the establishment of a vision. This can be achieved not only at the individual level but also at the organizational level. When an individual's work performance is based on these core ethical values, it creates a solid framework for the long-term and sustainable creation and delivery of trust to all stakeholders.

4.4.1 Approach

Ethical excellence is enabled through an iterative operation of concurrent steps shaping, a spiral form of the "*3E's*" as illustrated in Fig. 4.3. This structure facilitates the reasonable accomplishment of optimal results.

1. *Explore*: an exceptional quality search for the core ethical values that are the foundation for delivering excellence in all areas.
2. *Envision*: using the process of disciplined imagination to visualize the establishment of ethical standards to create sustainable success in the future.
3. *Execute*: develop a strategic action plan to translate the ethical vision into real practice.

4.4.2 Process

Ethical excellence does not come about in isolation, but rather a number of processes come together to produce it. In this case, the processes of *exploration*, *envision*, and *execution* come together to produce ethical excellence.

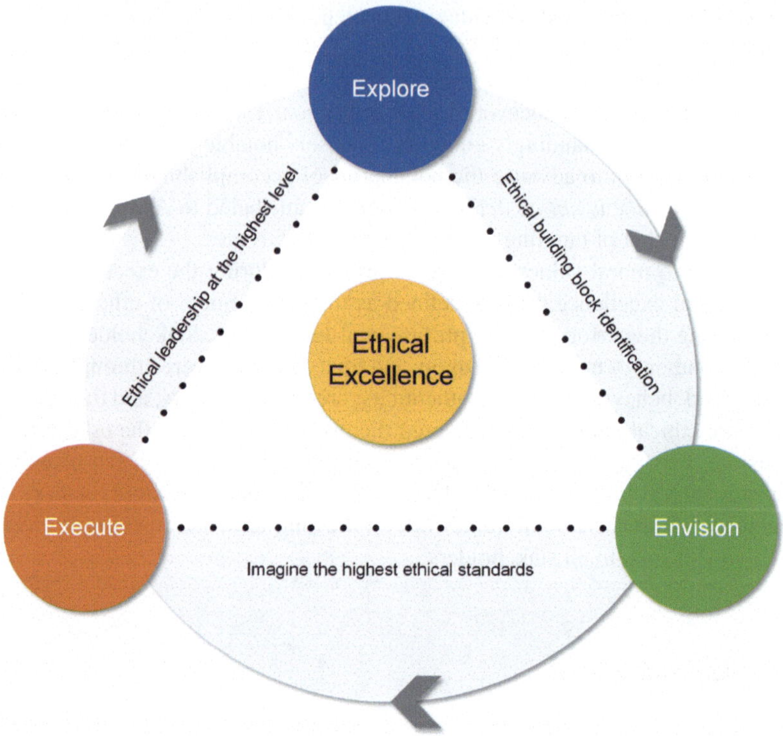

Fig. 4.3 The 3E's of ethical excellence approach

Explore

The process is born of an external creative impulse and begins with an inspired idea to give a concrete form to the core ethical values of the organization. These values are shaped by our professional heritage and the experiences of our careers. It is a matter of recognizing the most excellent and effective "*what is*" by focusing on our most significant past experiences in relation to the upholding of ethical values.

There are several approaches that can be used to facilitate this sensitive phase and to achieve positive outcomes. Various types of appreciative discussions, reflective conversations, and the sharing of positive experiences can generate enthusiasm and joy.

The expected outcome of this phase is a set of deeply held ethical values that have been the cornerstone of one's life and professional success.

Envision

During this phase, we are trying to form a mental picture, to visualize and contemplate the actual events that are taking place in the external world. Our internal beliefs, feelings, and expectations are the blueprints for these actual events to occur.

In this stage, we appropriately use disciplined imagination to form new ideas, images, or concepts of external objects that are not present to the senses. We target specific ethical behaviors, events, or outcomes that we expect to develop in our professional lives. On a personal level, we develop a perception of ethical standards and establish our standards for a sustainable professional future.

In addition, paired conversations between people about the possibilities of the future, constructive and creative visualization, and the creation of strong images in the mind—or even allegorical images of the future state—are beneficial.

Execute

It is at this stage that the ethical vision is expressed as a set of ethical ideas that are clearly and precisely articulated. It may even be developed in a systematic way as an individual code of ethics, accompanied by a list of activities that can be successfully carried out. The implementation of the code of ethics requires a specific time frame.

Simulated processes and activities that are conducted for this specific purpose can facilitate the practice of resolution of ethical dilemmas. This is useful for building the skills needed to find solutions in complex situations that require a balance of ethical ideals and business results.

References

Ackrill, J. 1981. *Aristotle the philosopher*, 1st ed., 170. Oxford University Press. ISBN: 978-0192891181.

Aristotle. 2019. Nicomachean ethics. Trans. by T. Irwin. 3rd ed., 480. Indianapolis: Hackett Publishing Company, Inc. ISBN: 978-1624668159.

Bentham, J. 2007. *An introduction to the principles of morals and legislation*, 416. Dover Publications. ISBN: 978-0486454528.

Boatright, J., J. Smith, and B. Patra. 2017. *Ethics and the conduct of business*, 8th ed., 520. India: Pearson. ISBN: 978-9352862306.

Crane, A., and D. Matten. 2010. *Business ethics: managing corporate citizenship and sustainability in the age of globalization*, 3rd ed., 648. Oxford University Press. ISBN: 978-0199564330.

Ferrell, O., J. Fraedrich, and L. Ferrell. 2021. *Business ethics: ethical decision making and cases*, 13th ed., 640. Cengage Learning. ISBN: 978-0357513361.

Floridi, L., et al. 2018. AI4People'an ethical framework for a good AI society: opportunities, risks, principles, and recommendations. *Minds and Machines* 28: 689–707. https://doi.org/10.1007/s11023-018-9482-5.

Hartman, L., J. DesJardins, and C. MacDonald. 2024. *Business ethics: decision making for personal integrity and social responsibility*, 6th ed. McGraw Hill. ISBN: 978-1260260496.

Jennings, M. 2017. *Business ethics: case studies and selected readings*, 9th ed., 416. Cengage Learning. ISBN: 978-1305972544.

Kidder, R. 2009. *How good people make tough choices* Rev Ed: Resolving the Dilemmas of Ethical Living, 272. Harper Perennial. ISBN: 978-0061743993.

MacIntyre, A. 2016. *After virtue: a study in moral theory*, 3rd ed., 286. University of Notre Dame Press. ISBN: 978-0268035044.

Mittelstadt, B., et al. 2016. The ethics of algorithms: Mapping the debate. *Big Data and Society* 3 (2): 1–21. https://doi.org/10.1177/2053951716679679.

Pakaluk, M. 2005. *Aristotle's Nicomachean ethics: an introduction*, 360. Cambridge University Press. ISBN: 978-0521520683.

Rest, J. 1986. *Moral development: advances in research and theory*, 3rd ed., 241. Praeger. ISBN: 978-0275922542.

Rhode, D. 2013. *Lawyers as leaders*, 1st ed., 320. Oxford University Press. ISBN: 978-0199896226.

Schwartz, P., and D. Solove. 2011. The PII problem: privacy and a new concept of personally identifiable information. *New York University Law Review* 86: 1814.

Shaw, W. 2016. *Business ethics: a textbook with cases*, 9th ed., 512. Cengage Learning. ISBN: 978-1305582088.

Shields, C. 2014. *Aristotle*, 2nd ed., 528. Routledge. ISBN: 978-0415622493.

Solomon, R. 1993. *Ethics and excellence: cooperation and integrity in business*, 304. Oxford University Press. ISBN: 978-0195087116.

Solove, D. 2010. *Understanding privacy*, 277. Harvard University Press. ISBN: 978-0674035072.

Trevino, L., and K. Nelson. 2021. *Managing business ethics: straight talk about how to do it right*, 8th ed., 348. Wiley. ISBN: 978-1119718857.

Vallor, S. 2016. *Technology and the virtues: a philosophical guide to a future worth wanting*, 328. Oxford University Press. ISBN: 978-0190498511.

Chapter 5
Ethics and Responsibility

Abstract The chapter on ethics and responsibility examines the ethical dimensions and responsibilities associated with the recording and documentation of cultural heritage. It emphasizes the importance of preserving cultural heritage for future generations while addressing the ethical and moral considerations that shape this practice. The chapter traces the historical context of cultural heritage documentation, highlighting past biases and colonial influences, and underscores the shift towards more inclusive and community-involved approaches. Key ethical principles include respecting community perspectives, obtaining informed consent, and maintaining cultural sensitivity. The chapter discusses the challenges of access, ownership, and the potential biases in documentation, along with the ethical dilemmas posed by modern digital technologies. Practical obligations for professionals are outlined, such as producing high-quality records, raising awareness, ensuring smooth information transfer, and maintaining transparency in workflows. The importance of knowledge transfer, capacity building, participation in professional networks, and adherence to international standards is emphasized to uphold ethical practices and foster global collaboration in cultural heritage preservation.

5.1 Introduction

The term cultural heritage encompasses the physical artifacts and intangible attributes of a group or society that have been inherited from past generations and that have historical, social, and aesthetic significance. The recording and documentation of these assets ensures that their value will be preserved for future generations and enables societies to maintain a link with their past. However, this process is not merely technical. It is deeply intertwined with ethical and moral considerations that shape how cultural heritage is understood and valued (Smith 2007). The ethical aspects of recording and documenting cultural heritage and the responsibilities of those involved in this work are explored in this chapter.

The recording and documentation of the physical characteristics of historic structures and sites form the basis for preventive maintenance, monitoring, and preservation, as well as for archival and educational purposes. The result of such campaigns

E. Stylianidis, *Exploring the Ethical Dimension in Recording and Documenting Cultural Heritage*, https://doi.org/10.1007/978-3-031-80034-4_5

is the collection and production of a considerable amount of information. This information can be used to make decisions about conservation actions. It can also serve as material for training activities, and for the education of younger generations of professionals in the field of cultural heritage. Last but not least, a detailed recording and documentation campaign can also become the main resource through which we can "revisit" a historic structure or a site that has been drastically altered or has disappeared due to natural or human-made disasters.

Technology is revolutionizing the field of heritage preservation by expanding our ability to record and document. 3D imaging techniques such as digital photogrammetry (captured reality) and 3D computer modeling allow for detailed preservation of both tangible and intangible heritage (Carvajal et al. 2020). However, these advances have also brought new ethical dilemmas related to accessibility, ownership, and the representation of marginalized voices. The challenge is how to use these tools responsibly while ensuring that cultural heritage is preserved in a way that honors its intrinsic values and significance.

A fundamental principle in the recording and documentation of cultural heritage is the respect for the communities from which the heritage comes. The recording and documentation of cultural heritage should include the informed consent of local communities, especially in indigenous and minority contexts (Logan 2007). Without this respect, documentation efforts run the risk of imposing external interpretations and potentially misrepresenting or misappropriating cultural narratives. This can lead to the distortion of history and identity. Ensuring that these practices are inclusive and culturally sensitive is the responsibility of researchers and practitioners.

Ethical guidelines for documenting cultural heritage have been established at the international level, including the UNESCO Conventions and the ICOMOS Charter, which emphasize the importance of ethical standards in practice. However, these guidelines need to be translated into context-specific protocols that take into account the different ways in which communities perceive and interact with their cultural heritage.

The recording and documentation of cultural heritage is a complex field that requires a careful balance to be struck between the preservation of cultural heritage and ethical responsibility. The way we document cultural heritage will continue to be shaped by the interplay between evolving technology and ethical principles.

5.2 Historical Context

The history of cultural heritage recording and documentation is a reflection of the evolution of attitudes toward culture, knowledge, and preservation. Early efforts at documentation were often undertaken by explorers, archaeologists, and missionaries. In their quest for discovery and understanding, they collected artifacts and created written accounts. While these efforts were groundbreaking, they were often ethnocentric and steeped in colonial bias, resulting in the appropriation of cultural heritage without proper consideration of the perspectives of the communities from which it originated (Meskell 2005).

During the nineteenth century and the first half twentieth century, collecting and documenting cultural heritage frequently served colonial and Western museum interests. Objects and artifacts were often removed from their cultural contexts and transported to distant countries, primarily for display in museums in Europe and the United States (Knowles and Gosden 2001). This has led to the creation of significant collections that are often lacking in comprehensive contextual information, which poses a challenge to their accurate interpretation today.

In the mid-20th century, coinciding with the rise of cultural heritage as a field of study, there was a shift toward more scientific and systematic methods of documentation. The global recognition of the importance of cultural heritage, the need for its preservation, and the ethical implications of such work were marked by the establishment of UNESCO in 1945 and its subsequent conventions. The Venice Charter (1964) also played a crucial role in establishing ethical guidelines for the conservation and restoration of monuments and sites. This influenced the approach to documentation in the following decades.

In the late twentieth century, there was a significant increase in awareness of the need to involve local communities and indigenous groups in the documentation of cultural heritage. This shift was influenced in part by post-colonial critique, which highlighted the ways in which previous documentation practices had marginalized the voices of indigenous peoples and distorted historical narratives (Smith 2007). The adoption of the Nara Document on Authenticity (1994) was a reflection of this shift, emphasizing the diversity of cultural expressions and the importance of context in the understanding of authenticity.

A heightened sense of ethical responsibility, technological advances, and a deeper understanding of the complexities of cultural significance are shaping the documentation of cultural heritage today. The digital age has opened up new possibilities for recording and sharing cultural heritage. However, it has also introduced challenges related to data ownership, representation, and access. As a result, contemporary approaches to documentation must navigate a complex landscape in which historical lessons are at the heart of modern practices.

5.2.1 *Principles of Ethics and Responsibility*

It is a fundamental ethical principle to respect the communities that own, care for, or value cultural heritage. Documentation should involve the meaningful participation of these communities so that their voices can shape the representation and understanding of their heritage (Logan 2007). This principle requires researchers to approach documentation with humility. They must recognize their role as facilitators rather than arbiters of cultural knowledge. Engaging with communities builds trust and ensures that the resulting documentation reflects the values and perspectives of the community.

When documenting cultural heritage, obtaining informed consent is crucial. This is the process of obtaining explicit permission from the community or organization representatives before the documentation of their cultural property. The consent process should include clear communication about how the documentation will be used, how it will be stored, and how it will be shared. In addition, intellectual property rights must be respected. This is especially true in the digital age where content can be easily distributed. Additional care must be taken to protect these intangible assets, as traditional knowledge and expressions often lack formal legal protection.

Understanding the cultural significance of heritage and how to present it in an appropriate manner is essential to becoming culturally sensitive. Cultural sensitivity requires an awareness of the symbolic meanings, rituals, and practices that are associated with the objects or sites of cultural heritage (Smith 2007). Misinterpretation can lead to the erosion of the cultural significance of heritage or to the unintentional offending of communities. Researchers should approach documentation with sensitivity to cultural nuances and avoid imposing their own biases. This will ensure that the resulting records accurately reflect the significance of the heritage.

There are significant ethical challenges associated with issues of access and ownership. Ownership of cultural heritage is often a matter of dispute, especially in the case of artifacts that have been removed from their original contexts. Documentation should be cognizant of these disputes and should strive to present a balanced view that reflects a variety of perspectives (O'Keefe and Prott 1992). In addition, in the digital age, where the ease of sharing can conflict with the desire to protect cultural assets from exploitation, the issue of access becomes more complex. Therefore, balancing the need for access with respect for property rights should be the goal of ethical documentation practices.

Researchers need to be mindful of the biases that can emerge during documentation. Historical biases in the interpretation of cultural heritage have often had the effect of marginalizing minority voices and the creation of distorted narratives (Smith 2007). Ethical documentation practices should strive to counter these biases by adopting an inclusive approach that seeks out and amplifies the perspectives of those who are underrepresented. This requires a proactive effort to understand the broader context of the heritage and to ensure that documentation practices are reflective of a diverse range of voices.

5.2.2 Challenges and Dilemmas

A dilemma is a situation in which a difficult choice must be made between two or more alternatives, especially equally unpleasant ones. In cultural heritage recording and documentation, the ethical dilemma can be summarized by the following general question:

Can we apply ethics to the digital recording and documentation of cultural heritage to ensure its preservation?

Professionals dedicated to the recording and documentation of cultural heritage should always incorporate ethics into their daily practices. There are several important ethical considerations that must be addressed during these activities.

One of the major challenges in recording and documenting cultural heritage is access and ownership issues. These complexities are often magnified in cases where artifacts have been removed from their original cultural contexts. This can lead to contentious debates about their rightful ownership and the ethics of displaying them in foreign institutions. For example, many artifacts held in European museums are the subject of calls for repatriation to their countries of origin. This raises questions of legal rights, historical justice, and cultural reconciliation (Greenfield 2007). Ethical documentation must navigate these delicate issues. It must strive to respect the perspectives and rights of the communities originally associated with these artifacts.

Unprecedented opportunities for the preservation and dissemination of cultural heritage have arisen with the advent of digital technologies. But the transformation brings significant challenges, especially around data ownership and the potential for digital colonialism. As cultural heritage materials are digitized, questions arise as to who will control the digital representations (Pepe et al. 2024) and how they will be accessed. Digital decay and rapid technological obsolescence further complicate long-term preservation efforts. Ethical practices therefore need to ensure that digital projects enhance, rather than exploit, the cultural heritage they are intended to preserve, and that they do so in a way that is sustainable and inclusive.

Documentation processes are by their very nature shaped by the perspectives and biases of those who carry them out. The result can be biased representations that lead to the marginalization of certain groups or the omission of important aspects of cultural narratives. Historical documentation efforts have often reflected the priorities and biases of dominant cultures. As a result, minority and indigenous perspectives have been underrepresented (Smith 2007). In order to address these biases and ensure that cultural heritage documentation is comprehensive and representative, conscious efforts must be made to include diverse perspectives and redress historical imbalances.

Political contexts can also strongly influence the documentation of cultural heritage. The politics of culture acknowledges "culture" as an important locus of political contestation over meaning, where various social groups are negotiating issues of identity, belonging, inclusion, and exclusion (Korf 2020). Governments and other powerful entities may use cultural heritage as a tool for the forging of national identities, potentially skewing documentation toward narratives that support current political agendas or ideological stances. Such influences have the potential to compromise the integrity of documentation efforts, making it imperative that ethical standards and scholarly independence be maintained in the face of political pressures.

Ethical decision-making is a constant in the process of cultural heritage documentation. There are ethical implications to decisions about what to document, how to interpret it, and who to involve. These decisions can have a significant impact on how the cultural heritage will be understood and valued by current and future generations. Navigating these decisions requires a robust ethical framework and a commitment to transparency, inclusiveness, and respecting cultural diversity (Logan 2007).

5.2.3 Practical Obligations

The recording and documentation of cultural heritage is not merely an academic exercise. It carries with it profound ethical and practical obligations (Fig. 5.1) that must be rigorously adhered to in order to respect the cultural and historical significance of the objects. This ensures that the documentation process will be conducted in a responsible manner that respects the integrity of the cultural property and the community that owns it. Through adherence to a structured set of practical obligations, practitioners can maintain the highest standards of cultural sensitivity, legal compliance, and professional integrity. This framework is essential to the fostering of trust and cooperation between cultural heritage professionals and community stakeholders so that the legacy of cultural heritage can be preserved and respected for future generations.

Production of High-Quality Records

Professionals involved in recording and documenting campaigns collect and generate numerous electronic files, usually in heterogeneous formats such as images, drawings, and point clouds (Fig. 5.2). All of these files should not only be of high

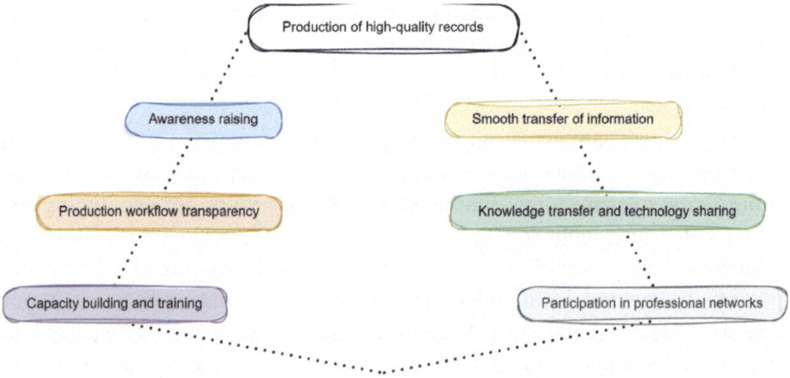

Fig. 5.1 Practical obligations to ensure responsibility in cultural heritage recording and documentation

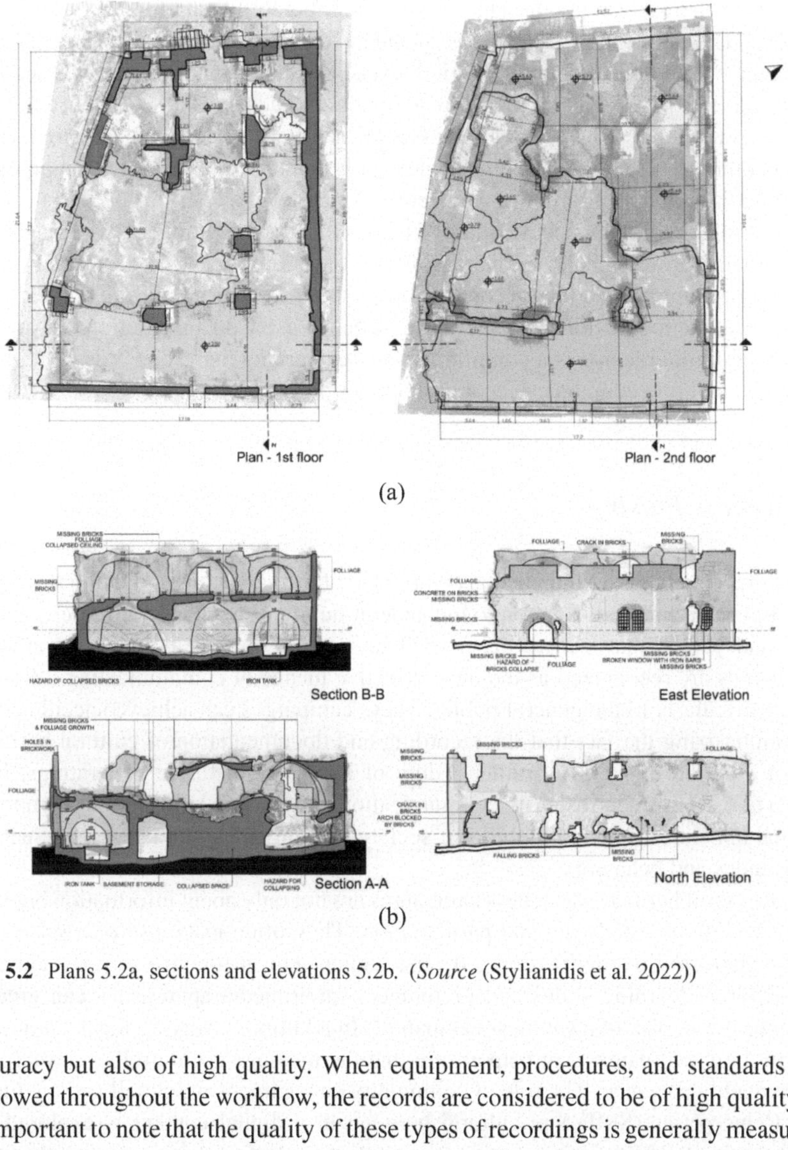

Fig. 5.2 Plans 5.2a, sections and elevations 5.2b. (*Source* (Stylianidis et al. 2022))

accuracy but also of high quality. When equipment, procedures, and standards are followed throughout the workflow, the records are considered to be of high quality. It is important to note that the quality of these types of recordings is generally measured in an objective manner, usually on the basis of specific metrics and statistical analysis. In a study funded by the European Commission, a report has been produced that details the imaging parameters, formats, standards, benchmarks, methodologies and guidelines related to the quality of 3D digitization of tangible cultural heritage (Commission, European, Content Directorate-General for Communications Networks, and Technology 2022).

Adherence to established digital preservation standards and guidelines is essential for ensuring the highest quality records creation. Using standardized formats and metadata schemas can greatly enhance the interoperability and longevity of digital records. The application of rigorous metadata standards is critical to the effective management and preservation of digital assets. Digital resources are represented by metadata, and trust lies in proving their authenticity (Pacheco et al. 2023). In addition, the integration of quality control processes at each stage of document creation, from data capture to file storage, ensures that records meet required quality thresholds.

In addition, ongoing training and professional development in the latest digital recording technologies and standards is essential for practitioners. Ongoing education helps professionals stay abreast of the latest advances and best practices in digital recording, which in turn contributes to the production of high-quality records. In essence, the foundation for achieving excellence in the documentation and preservation of digital records is a commitment to maintaining high standards in both the technical and procedural aspects of digital recording.

Awareness Raising

Awareness raising is a fundamental component of primary communication strategies that aim to provide information to the general public or to change attitudes, behaviors, and beliefs that lead to the normalization of a particular image. Awareness campaigns are recognized as the most effective means of communicating information, particularly to the general public. These campaigns can achieve specific goals by emphasizing the fact that the recording and documentation of cultural heritage is not a private matter, but rather a duty of humanity to future generations. It is also a two-way street, fostering communication and sharing information to improve mutual understanding and mobilizing society to bring about the necessary changes in attitudes and behaviors.

Successful heritage awareness campaigns are not only about information but also about community dialogue and participation. They often make use of a variety of media platforms in order to reach a diverse audience in an effective way. By addressing different learning styles and preferences, multimedia approaches can greatly enhance the impact of awareness campaigns. In addition, involving local communities in these campaigns can help ensure that the message is culturally relevant and more readily accepted. The role of community engagement and Small and medium-sized enterprises (SMEs) in cultural heritage in a digital context is explored by (Murphy et al. 2023).

There is also evidence that the strategic use of social media can have a positive impact on the reach and effectiveness of awareness-raising campaigns. The use of social media has become an integral part of the lives of many people around the world. Social and digital marketing offers significant opportunities for organizations to reduce costs, increase brand awareness, and drive sales (Dwivedi et al 2021). Social media platforms allow for rapid dissemination of information and facilitate interactive engagement with audiences. They provide a dynamic space for discussion and feedback. Social media platforms have the potential to bridge the gap between different groups in the community and to foster broader engagement and participation in heritage conservation. Incorporating these digital tools into awareness-raising campaigns can therefore make a significant contribution to successfully communicating messages and achieving campaign goals. 3D digital technologies for community engagement through co-creation can explore how the meaning of a site can be influenced and challenged through community recording (Jeffrey et al 2020).

Smooth Transfer of Information

The smooth transfer of information is a key factor in maintaining the continuity and integrity of knowledge across different platforms and between different stakeholders. The effective transfer of information is not only about the dissemination of the knowledge itself; it is also about the accurate and timely delivery of information in a way that is accessible and understandable to the target audience. The use of transparent and accessible information transfer workflows provides the opportunity for all stakeholders to have a common understanding and be able to apply the information effectively in their context.

In the digital era, technology significantly enhances the efficiency of information transfer. Cloud-based storage solutions and information systems are critical to improving data management and distribution workflows. These technologies help mitigate the formation of data silos by centralizing data repositories, ensuring that all authorized users have access to the latest information, which is essential for effective decision-making—strengthening at the same time transparency and seamless information flow. Integrating advanced analytics tools into these systems can provide real-time insights, which are crucial for decision-makers to interpret complex datasets quickly and accurately. This approach not only facilitates access to data but also streamlines organizational processes, leading to more informed decisions and improved operational efficiency.

Smooth information transfer requires a blend of technology solutions, standardized practices, and skilled personnel. By investing in the right technologies, adopting standardized communication protocols, and fostering continuous professional development, professionals and organizations can achieve more efficient and error-free information transfer. Ultimately, this improves operational effectiveness and stakeholder engagement.

Production Workflow Transparency

Transparency in the production workflow is essential for the effective documentation and preservation of cultural heritage. This includes the clear documentation of all processes, from the collection of data to the archival storage of the records. This openness not only ensures the authenticity and accuracy of the historical record, but it also increases the level of trust among stakeholders, including researchers, heritage professionals, and the general public. Transparent workflows allow for better oversight and accountability, which are critical to maintaining the integrity of cultural heritage recording and documentation efforts.

In the field of cultural heritage, transparency is achieved through the implementation of standardized procedures for the collection and reporting of data. For example, the use of universally accepted metadata standards and documentation guidelines, as recommended by the International Council on Archives (ICA), ensures that the data collected and preserved can be easily understood and used by professionals across regions and disciplines. (Skinner 2014) provides an overview of the literature related to metadata in the context of archives and museums, while outlining best practices, theories, and models in this field. In addition, transparency can be further enhanced by providing a traceable and auditable trail of data modifications and access through the use of digital tools such as digital asset management systems.

Furthermore, the public engagement associated with transparency should not be underestimated. Institutions foster greater appreciation and understanding of cultural heritage by making the documentation process accessible to the public. This can include providing online access to digital archives. It can also include creating interactive platforms where the public can observe and even participate in the documentation process. Engaging the community in this way not only educates and informs, but also fosters a collective responsibility to preserve heritage. (McCarthy 2024) presents multi-image photogrammetry as a practical tool for heritage survey and community engagement. He highlights its potential as a practical, everyday tool for archaeological practitioners to use in two of the main types of commissioned archaeological work: rapid survey and community engagement.

A commitment to transparency in the production of cultural heritage documentation is fundamental to preserving the authenticity and utility of historical records. It ensures that these treasures of human history will not only be preserved but that they will also be respected and understood by future generations.

Knowledge Transfer and Technology Sharing

Knowledge transfer is the act of sharing skills, information, expertise, or ideas with others. It can involve a variety of methods. These include mentoring, training, coaching, or simply sharing information through communication channels. Sharing knowledge across an organization can encourage employees to take advantage of technological innovations. It can also increase productivity and efficiency. The goal of knowledge transfer is the sharing and retention of critical knowledge within an organization, team, or community. By sharing their expertise and experience, individuals are able to build on their successes and avoid the mistakes of others, leading to better decision-making and improved performance. A knowledge transfer system is a powerful tool that enables knowledge transfer to occur smoothly. This tool should be software, not tons of files and folders. It should allow you to easily grant access to your employees, customize the application, keep up with the latest technological changes, and track the knowledge transfer process.

Knowledge transfer and technology sharing are critical in the cultural heritage field. A knowledge management system aims to overcome coordination problems for knowledge sharing within and between cultural heritage projects (Nova et al. 2023). This makes valuable information and advanced methodologies accessible across regions and institutions. These processes not only democratize access to innovative tools, they also foster global collaborations that improve the quality and reach of cultural heritage conservation. Promoting cultural diversity and inclusiveness, the effective transfer of knowledge and technology across borders can empower local communities by providing them with the tools and training they need to preserve their own heritage.

Effective knowledge transfer requires careful planning, clear communication, and a structured approach, as shown in the Fig. 5.3. Here are some of the key steps to take into consideration when transferring knowledge:

Identification of knowledge to be transferred: Identifying what knowledge needs to be shared is the first step to effective knowledge transfer. It can be used to identify the skills, knowledge, and expertise needed to make a job, process, or role a success. For example, when an organization implements a new technology platform, the knowledge to transfer may include understanding how to use the platform, how to troubleshoot common problems, and how to use its features to improve productivity.

Identification of target audience: Once the knowledge has been identified, the next step is to determine who will be the recipients of the knowledge. Different people learn and process information in different ways, and knowledge transfer is not a one-size-fits-all process. In order to transfer knowledge effectively, it is important to tailor the approach to the needs and learning styles of the audience. This may involve adjusting the pace of the training. It may involve providing hands-on experience or using real-world examples. On this basis, you can decide

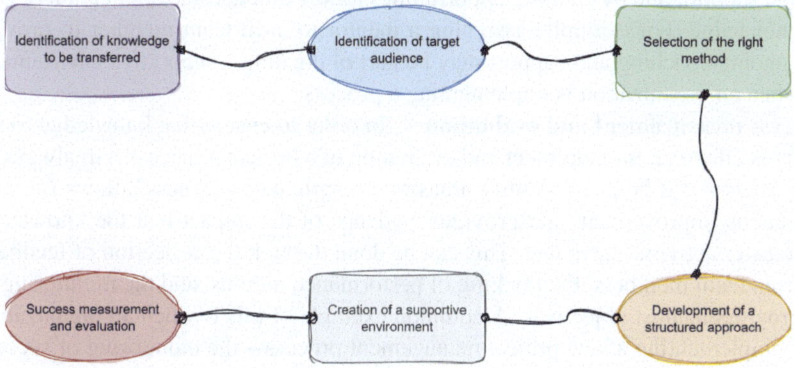

Fig. 5.3 Effective knowledge transfer workflow

whether a catchy phrase is more appropriate for the sharing of knowledge than more formal methods.

Selection of the right method: For some people, the best way to learn is through hands-on experience, while others may prefer to learn by reading or listening to lectures. Therefore, it is important to choose the right method of knowledge transfer accordingly. There are many methods that can be used to transfer knowledge. These include training, coaching, mentoring, documentation, and job shadowing. By choosing the appropriate method of knowledge transfer, you can ensure that the information is presented in a way that is engaging, accessible, and effective for the target audience. This can lead to a better retention of the knowledge being shared and a better application of the skills being taught. For example, if an organization is training employees on a new software program, a hands-on training session may be more effective than documentation.

Development of a structured approach: The development of a structured approach to knowledge transfer is important because it will ensure that the process is organized, consistent, and effective. Without a structured approach, knowledge transfer can be haphazard and inconsistent, which can lead to miscommunication, errors, and inefficiencies. In order to ensure that the knowledge transfer process is effective and efficient, it is important to develop a structured knowledge transfer checklist that includes clear goals, timelines, and milestones. This can help to keep the process on track and ensure that the most important knowledge is being transferred in a timely manner. As an example, when launching a new product or service, the plan may include a 4-week period of systematic training to ensure that all team members receive adequate training and coordination.

Creation of a supportive environment: If you want to get your message across, it is important to set up an environment that is conducive to learning. The effective transfer of knowledge is crucial because it boosts morale, makes employees happier, and makes it easier for them to learn new things and to stay with the company for a longer period of time. Organizational success and information sharing can also be enhanced by fostering an atmosphere of encouragement. This can be done by encouraging team members to ask questions, by providing feedback and support, and by creating opportunities for the practice and application of new knowledge. For example, assigning a mentor to each team member to provide ongoing coaching and support may be part of creating a supportive environment when an organization is implementing a process.

Success measurement and evaluation: In order to ensure that knowledge transfer is effective, measurement and evaluation of a project's success will always be a necessity. It helps to identify areas of strength and weakness, allows for continuous improvement, and provides evidence of the impact that the knowledge transfer activities have had. This can be done through the collection of feedback from team members, the tracking of performance metrics, and the monitoring of progress against objectives. A common example of this is when an organization is implementing a new project management process—the monitoring of success may include the tracking of project schedules and budgets and collecting feedback from team members on the effectiveness of the new process.

Technology sharing in the cultural heritage field often involves the use of digital tools, such as 3D scanning and virtual reality, for the documentation and interpretation of heritage sites and artifacts (Barszcz et al. 2023). These technologies not only help create detailed and accurate records, but also make these records more accessible to the global community. Integrating these advanced digital technologies into cultural heritage documentation practices allows for a more nuanced understanding of cultural sites and facilitates their virtual preservation. This is particularly valuable in contexts where physical preservation is challenging.

Effective knowledge transfer also requires structured training programs and workshops that expose heritage professionals to the latest methodologies and technologies. Partnerships between academic institutions, international organizations, and local communities play a crucial role in this regard. For example, CIPA—Heritage Documentation, the dynamic international organization that has the dual responsibility of keeping pace with technology and ensuring its usefulness for cultural heritage conservation, education, and dissemination—frequently conducts training sessions on best practices in digital documentation, with the aim of standardizing skills on a global scale while adapting them to local contexts and needs.

To build the capacity of cultural heritage professionals worldwide, knowledge transfer and technology sharing are essential. These efforts will ensure that the benefits of technological advances are widely distributed and culturally sensitive, in addition to contributing to more equitable and sustainable preservation of the world's heritage.

Capacity Building and Training

Capacity building and training are essential obligations for all professionals and organizations involved in the recording and documentation of cultural heritage. These practices ensure that heritage sites are carefully preserved and accurately documented. It is also a prerequisite for the promotion of sustainable conservation practices. ICOMOS and all international organizations and associations emphasize the need for ongoing training programs to keep heritage professionals abreast of the latest technologies and methodologies that are critical to effective heritage conservation.

Capacity building involves enhancing the skills, knowledge and competencies of individuals and organizations involved in cultural heritage documentation. This includes both formal education and on-the-job training tailored to specific needs and challenges. For example, in regions rich in archaeological sites, training might focus on advanced archaeological survey techniques using different types of sensors, as well as the use of Geographic Information Systems (GIS) for mapping and analysis. Such specialized training will ensure that cultural heritage professionals are able to use the best practices and tools available, and thus help to preserve the integrity and authenticity of cultural monuments, objects and sites.

In addition to technical skills, capacity building must include an understanding of legal and ethical considerations. International conventions, such as the UNESCO

World Heritage Convention, as well as national laws that govern the preservation of cultural heritage, need to be well understood by heritage professionals. Modules on legal frameworks, ethical standards, and the importance of community engagement should therefore be included in training programs. This holistic approach will ensure that all aspects of the documentation of cultural heritage are handled with due diligence and with respect for cultural sensitivities (UNESCO 2013).

In addition, capacity building extends to the development of institutional frameworks that support heritage conservation. This includes the establishment of robust documentation systems, the creation of databases for heritage records, and the promotion of interdisciplinary collaboration. Institutions will also need to invest in ongoing professional development programs to keep their staff abreast of emerging trends and new technologies. By fostering a culture of lifelong learning, cultural heritage organizations are better equipped for changing challenges and opportunities in their fields (Smith 2007).

The importance of international cooperation and the exchange of knowledge in the building of capacity cannot be overemphasized. Participation in international workshops, conferences, and exchange programs enables cultural heritage professionals to learn from global best practices and adapt them to their local contexts. Organizations such as ICOMOS and UNESCO have a key role to play in facilitating this exchange and in promoting a global standard for the documentation and conservation of cultural heritage. Collaborative efforts ensure that knowledge and skills are not confined to one region. Instead, they are shared widely to the benefit of the global heritage community.

Participation in Professional Networks

Participation in professional networks is an essential obligation for all those involved in the recording and documentation of cultural heritage. These networks provide a platform for the exchange of knowledge, collaboration, and dissemination of best practices, all of which are essential for the effective conservation of cultural heritage. ICOMOS and many other relevant international organizations and committees encourage their members to participate in professional networks. This helps heritage professionals keep abreast of the latest developments in the field and fosters a community of practice that supports continuous improvement.

Professional networks offer many opportunities for education and professional development. Through conferences, workshops, and seminars, cultural heritage professionals can enhance their skills and learn about innovative techniques and technologies. For example, CIPA—Heritage Documentation regularly organizes annual summer schools to provide hands-on training in recording and documentation practices. In addition, the International Centre for the Study of the Preservation and Restoration of Cultural Property (ICCROM) regularly organizes training programs and workshops on various aspects of heritage conservation. These training activities facilitate the exchange of experience and knowledge among practitioners from different regions and disciplines, thereby enriching their understanding and approaches to heritage documentation.

In addition to providing formal opportunities for continuing education, professional networks also offer informal systems of support. It is through these networks that members can seek advice, share challenges, and collaborate on projects. This sense of community is particularly valuable when dealing with complex cultural heritage issues that require a multi-disciplinary approach. For example, CIPA—Heritage Documentation provides a forum for professionals to work together to develop methods and tools for cultural heritage recording and documentation, ensuring that the practice is consistent and of high quality around the world.

Participating in professional networks also helps to advocate and raise the visibility of cultural heritage issues. By joining forces, cultural heritage professionals can be more effective in advocating for the policies and resources needed to preserve cultural heritage. Networks such as Europa Nostra play an important role in raising awareness of heritage sites that are under threat and in lobbying for the protection of these sites at the European level. Collective advocacy efforts can lead to greater recognition of the importance of cultural heritage and the allocation of the necessary funding and support.

International collaboration and exchange of best practices are facilitated by professional networks. Global collaborations that enable heritage professionals to learn from and apply knowledge from different cultural contexts are fostered by organizations such as UNESCO and ICOMOS. These international connections are critical to addressing the transnational challenges that cultural heritage faces, such as climate change and the illicit trafficking of cultural objects. Through their involvement in these networks, cultural heritage professionals are helping ensure that cultural heritage is preserved globally for future generations (UNESCO 2013).

References

Barszcz, M., et al. 2023. 3D scanning digital models for virtual museums. *Computer Animation and Virtual Worlds* 34 (3–4). https://doi.org/10.1002/cav.2154.

Carvajal, D., M. Morita, and G. Bilmes. 2020. Virtual museums. Captured reality and 3D modeling. *Journal of Cultural Heritage* 45: 234–239. https://doi.org/10.1016/j.culher.2020.04.013.

Commission, European, Content Directorate-General for Communications Networks, and Technology. 2022. Study on quality in 3D digitisation of tangible cultural heritage: mapping parameters, formats, standards, benchmarks, methodologies, and guidelines. VIGIE 2020/654—Final Study Report. https://doi.org/10.2759/471776.

Dwivedi, Y., et al. 2021. Setting the future of digital and social media marketing research: Perspectives and research propositions. *International Journal of Information Management* 59: 102168. https://doi.org/10.1016/j.ijinfomgt.2020.102168.

Greenfield, J. 2007. *The return of cultural treasures*, 3rd ed., 522. Cambridge University Press. ISBN: 978-0521802161.

Jeffrey, S., et al. 2020. 3D visualisation, communities and the production of significance. *International Journal of Heritage Studies* 26 (9): 885–900. https://doi.org/10.1080/13527258.2020.1731703.

Knowles, C., and C. Gosden. 2001. *Collecting colonialism: material culture and colonial change*, 256. Berg Publishers. ISBN: 978-1859734032.

Korf, B. 2020. Cultural politics. In *International encyclopedia of human geography*, ed. Audrey Kobayashi, 2nd ed., 111–116. Oxford: Elsevier. ISBN: 978-0081022962. https://doi.org/10.1016/B978-0-08-102295-5.10808-X.

Logan, W. 2007. *Cultural heritage and human rights*, ed. H. Silverman, and F. Ruggles. Closing Pandora's box: human rights conundrums in cultural heritage protection, 33–52. https://doi.org/10.1007/978-0-387-71313-7_2.

McCarthy, J. 2014. Multi-image photogrammetry as a practical tool for cultural heritage survey and community engagement. *Journal of Archaeological Science* 43: 175–185. https://doi.org/10.1016/j.jas.2014.01.010.

Meskell, L. 2005. *Archaeologies of materiality*, 229. Blackwell Publishing Ltd. ISBN: 978-1405136174.

Murphy, M., et al. 2023. Community Engagement in Cultural Heritage A Digital Context. Dublin. https://arrow.tudublin.ie/level3/vol18/iss1/1. Accessed 12 May 2024.

Nova, N., et al. 2023. A knowledge management system for sharing knowledge about cultural heritage projects. *Journal of Cultural Heritage* 63: 61–70. https://doi.org/10.1016/j.culher.2023.07.013.

O'Keefe, L., Prott P. 1992. Cultural heritage or cultural property? *International Journal of Cultural Property* 1 (2): 307–320.https://doi.org/10.1017/S094073919200033X.

Pacheco, A., C.G. Da Silva, and M.C.V. De Freitas. 2023. A metadata model for authenticity in digital archival descriptions. *Archival Science* 23: 629–673. https://doi.org/10.1007/s10502-023-09422-w.

Pepe, M., et al. 2024. Scan to BIM for the digital management and representation in 3D GIS environment of cultural heritage site. *Journal of Cultural Heritage* 50: 115–125. https://doi.org/10.1016/j.culher.2021.05.006.

Skinner, J. 2014. Metadata in archival and cultural heritage settings: A review of the literature. *Journal of Metadata Library* 14 (1): 52–68. https://doi.org/10.1080/19386389.2014.891892.

Smith, L. 2007. *Uses of heritage*, 368. Routledge. ISBN: 978-0415318310.

Stylianidis, E., et al. 2022. 3D documentation and visualization of cultural heritage buildings through the application of geospatial technologies. *Heritage* 5 (4): 2818–2832. https://doi.org/10.3390/heritage5040146.

UNESCO. 2013. Managing Cultural World Heritage. Paris. https://whc.unesco.org/en/managing-cultural-world-heritage/. Accessed 25 May 2024.

Chapter 6
Trustworthy Data

Abstract This chapter explores the importance of trustworthy data in the cultural heritage field, highlighting its critical role in the preservation and interpretation of historical legacies. Trustworthy data is the foundation for decision making, academic research, and public engagement. Challenges such as authenticity, degrading data, complexity, ethical considerations, and reliability are highlighted. Principles such as accuracy, completeness, consistency, security, and timeliness are fundamental to ensuring the trustworthiness of data. The chapter also discusses the importance of data collection standards, verification processes, and comprehensive documentation, as well as diverse data sources, including archaeological excavations, historical texts, oral histories, and technological advances. Ethical considerations and the challenges of maintaining the trustworthiness of data, such as technological limitations and human error, are also addressed. The chapter concludes with a discussion of the need for ongoing monitoring, updating, and staff training to ensure the reliability and preservation of cultural heritage data.

6.1 Introduction

The concept of trustworthy data is central to the field of cultural heritage, where the integrity, accuracy, and completeness of information are fundamental to the preservation and interpretation of the legacy of the past. As (Vecco 2010) succinctly defines it, cultural heritage is a collection of both tangible and intangible elements that have cultural significance. This broad scope underscores the complexity of the data collection process and the need to maintain high standards for the trustworthiness of the data. Trustworthy data is the foundation of all cultural heritage preservation efforts, from decision-making and academic research to public engagement and legal protection.

There are unique challenges to data trustworthiness in the cultural heritage domain. One of the most important concerns is the authenticity and veracity of the data that is collected. Cultural heritage objects and sites often have multi-layered histories and interpretations, which can vary widely depending on the source of information. As noted by (Jokilehto 2006), ensuring the authenticity of data in the cultural heritage field requires rigorous documentation practices and standards that need to

E. Stylianidis, *Exploring the Ethical Dimension in Recording and Documenting Cultural Heritage*, https://doi.org/10.1007/978-3-031-80034-4_6

be maintained consistently across different types of data and different methods of collection.

Another important issue to consider is the risk of degradation of the data over time. Cultural heritage data can be fragile whether it is ancient manuscripts susceptible to environmental degradation or digital data susceptible to technological obsolescence. Data longevity is critical, as emphasized by (Letellier 2007), who argue for robust information management systems to ensure that data are not only preserved, but also remain accessible and interpretable for future generations.

The complexity of the data in the cultural heritage field also poses a challenge. Data are collected from a variety of sources, including archaeological excavations, oral histories, and digital reconstructions. These data must be integrated and managed effectively. (Prechtel and Munster 2016) highlight the difficulties in the integration of traditional data with digital innovations, which often require new methodologies and tools for data interpretation and preservation.

Ethical considerations are also paramount in the collection and use of data related to cultural heritage. The ethical obligation to respect the rights of communities associated with cultural heritage artifacts and sites is critical to ensuring that data collection does not exploit or misrepresent the cultural significance of these sites. Furthermore, as (Smith 2007) points out, the involvement of local communities in the documentation process not only enhances the richness and accuracy of the data but also contributes to the preservation of the intangible aspects of cultural heritage.

The reliability of the data is a matter of constant concern. The methods used to collect, analyze, and store data must be carefully chosen to minimize the potential for error and bias. Trustworthy data is the cornerstone of the effective management and preservation of the cultural heritage. Addressing the challenges associated with authenticity, longevity, complexity, ethics, and reliability is essential to managing cultural heritage. Through adherence to established standards and continuous advancement of our methodologies, we can ensure the integrity and utility of cultural heritage data for present and future generations.

6.2 Principles of Trustworthy Data

Trustworthy data is essential for making effective decisions, conducting research, and operating across multiple sectors. It is a multi-dimensional concept that ensures the reliability, accuracy, and ethical use of data. Principles such as accuracy, completeness, consistency, security, and timeliness are at the core of trustworthy data (Fig. 6.1). Each of these principles contributes to the overall reliability and credibility of the data and makes it fit for the purpose for which it is intended.

Accuracy is one of the most critical aspects of trustworthy data. Accurate data is a true reflection of the real-world conditions or phenomena that it is intended to represent. The key to accuracy is the minimization of errors during data collection, entry, and processing. Inaccurate data can be the cause of incorrect analysis and

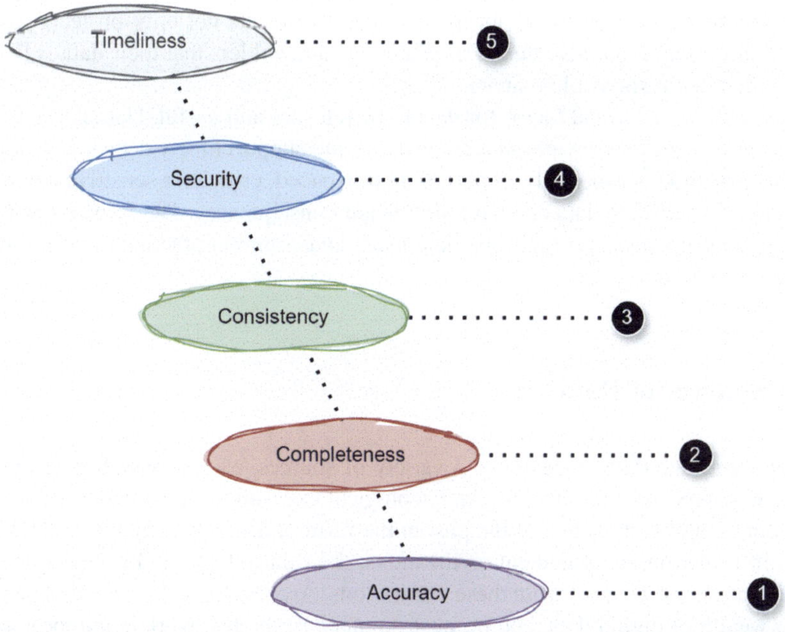

Fig. 6.1 Principles of trustworthy data

poor decision-making. Techniques such as validation and verification are often employed to increase the accuracy of the data. The cross-referencing of data entries with reliable sources or the use of automated checks can help to identify and correct inaccuracies.

Completeness refers to the extent to which all of the data you need are available. Data that is incomplete can be just as problematic as data that is inaccurate because it can lead to biased or partial analyses. The key to data completeness is the collection of all relevant data points and the filling in of gaps where possible. This may involve integrating data from multiple sources to provide a comprehensive view.

Consistency ensures that the data remain the same across different data sets and over time. Consistent data is critical for longitudinal studies and analyses that rely on multiple data sources to avoid inconsistencies and discrepancies. The establishment of data standards and formats can be a powerful driver of data consistency. We often make use of data frameworks to ensure that data collected by different actors or systems can be seamlessly integrated and compared to maintain consistency.

Security is essential to the protection of data from unauthorized access, loss, and manipulation. Both technical measures, such as encryption and access controls, and organizational policies, such as privacy laws and user training, are part of secure data practices. Ensuring the security of data not only protects sensitive information but also builds trust among stakeholders that their data is being handled in a responsible manner.

Timeliness is a critical factor for data to be relevant and useful. Data needs to be current in order to provide accurate insights and support timely decision-making. This principle is especially important in fast-paced, critical or sensitive environments. Out-of-date data can have significant consequences. The implementation of regular updates and real-time data feeds is a common practice to ensure that data is up to date.

6.3 Sources of Data

Cultural heritage data comes from a variety of sources, each of which contributes a unique perspective and insight. Archaeological excavations are a primary source, providing tangible evidence of the past in the form of artifacts, structures, and biofacts. Information about ancient civilizations, their daily lives, and socioeconomic conditions can be gleaned from these excavations. Archaeological excavation generates a wealth of digital data, and because so much of the fieldwork is unrepeatable, once the site is fully excavated, the digital records must be archived in a manner that best allows for reuse (Moody et al. 2021). However, the interpretation of archaeological data can be complex, often requiring careful consideration of context and sometimes limited by the extent and condition of the materials uncovered.

Historical texts serve as another important source of data for the cultural heritage. Documents provide direct accounts of historical events, societal norms, and individual experiences, such as manuscripts, official records, and personal letters. These texts can be the source of detailed narratives and chronological insights that are invaluable to the understanding of historical contexts. However, the reliability of historical texts must be subject to scrutiny, as they may be a reflection of the biases of their authors and the perspectives of particular groups or classes within society.

Oral histories are invaluable for preserving the intangible aspects of cultural heritage. They include traditions, languages, and personal memories. They capture the lived experiences and cultural expressions of communities and provide a rich and nuanced understanding of the past. By their very nature, however, oral histories are subjective and prone to the limitations of human memory, such as embellishment or forgetfulness. This subjectivity necessitates a careful approach to the corroboration of oral histories with other data sources in order to increase their reliability.

Advances in technology have led to the development of innovative methods for the capture and analysis of cultural heritage. Precise documentation of artifacts, monuments, and sites is possible with techniques such as photogrammetry (Fig. 6.2a) and 3D scanning (Fig. 6.2b). These digital tools facilitate the creation of accurate, high-resolution models that can be analyzed and shared on a global scale. While these methods offer unparalleled accuracy and preservation capabilities, they also require

(a)

(b)

Fig. 6.2 Recording and documentation with photogrammetry 6.2a and laser scanning 6.2b

significant technical expertise and resources, which can limit their accessibility (Alshawabkeh et al. 2022).

Each of these sources of data presents its own unique set of challenges related to reliability and validity. For example, the physical deterioration of artifacts can obscure critical details, while historical texts can omit or distort information based on the point of view of the author. Similarly, oral histories can be inconsistent and influenced by individual perceptions, even though they are rich in contextual detail.

Technological recordings, while accurate, rely heavily on the proper implementation and interpretation of the technology used. For a comprehensive understanding of cultural heritage, the integration of multiple data sources is essential. Researchers can cross-validate information and construct a more robust, multi-dimensional narrative of the past by combining archaeological evidence, historical texts, oral histories, and technological records. This holistic approach helps to mitigate the limitations of individual sources of information and increases the overall reliability and validity of data related to cultural heritage.

6.4 Data Collection Standards

Adherence to established standards and protocols during data collection is paramount to ensuring data trustworthiness. Comprehensive guidelines aimed at improving the accuracy, completeness, and documentation of data collection processes have been developed by organizations such as UNESCO and ICOMOS. These guidelines serve as a foundation for maintaining data integrity, which is critical for subsequent analysis and decision-making.

This includes the use of accurate instruments, standardized procedures, and trained personnel to minimize the risk of error. Accurate data collection ensures that the information collected is a true representation of the reality that it is intended to capture. In the context of cultural heritage, for example, accurate data about the condition and characteristics of a site are essential for its proper conservation and management. Errors in data collection can lead to misguided conservation efforts, potentially causing irreversible damage to cultural assets.

A study (Commission, Communications Networks, and Technology 2022) by the European Commission aims to promote 3D digitization across Europe. The study aims to identify parameters, formats, standards, benchmarks, methodologies, and guidelines for the 3D digitization of tangible cultural heritage. The overall objective is to promote the quality of 3D digitization projects by enabling cultural heritage professionals, institutions, content developers, stakeholders, and academics to define and produce high-quality digitization standards for tangible cultural heritage. The study identifies the key parameters of the digitization process and estimates their relative complexity and technological linkages, their effects on quality, and their different drivers. It also identifies 3D digitization standards and formats, including data types, formats, and metadata schemas for 3D structures.

International organizations also emphasize completeness. Incomplete data can lead to significant gaps in understanding and interpretation. Comprehensive data collection, covering all relevant variables and dimensions, is recommended by these organizations. In addition to physical measures, this should also cover contextual information including historical importance, current use, and environmental factors. Completeness allows researchers and practitioners to see an issue in its entirety and to make more informed and effective decisions.

Documentation is a key component of data collection. Proper documentation includes recording the methods used, the context of the data collection, and any potential biases or limitations that may have occurred. This transparency makes it possible for studies to be replicated, for results to be verified, and for potential sources of error to be identified. In the cultural heritage field, well-documented data can support future research and conservation efforts and provide a reliable reference for stakeholders.

Compliance with these standards also reduces the risk of data misinterpretation and loss of valuable information. Misinterpretation can occur if those analyzing the data do not clearly understand the methodology or if the data are taken out of context. By following established protocols, researchers ensure that data is presented in a clear and understandable manner. This reduces the likelihood of misinterpretation. In addition, thorough documentation and adherence to standards help protect data from loss, ensuring that valuable information is preserved for future use.

Critical to ensuring the trustworthiness of data collected in various fields, especially cultural heritage, are the standards and protocols established by international organizations. These guidelines emphasize the importance of accuracy, completeness, and documentation. Together, they contribute to the integrity and reliability of data. By adhering to these standards, researchers and practitioners can reduce the risks associated with data misinterpretation and loss, thereby improving the overall quality and utility of the data collected.

6.5 Data Verification

Verification processes play a critical role in ensuring the accuracy and reliability of the information collected after the diligent process of data collection in cultural heritage projects. This verification is essential not only for the integrity of the research but also for the preservation and interpretation of the cultural heritage property. Techniques such as cross-referencing with historical documents, corroboration of evidence from multiple sources, and the use of advanced technologies such as photogrammetry, laser scanning and GIS greatly enhance the reliability of heritage data.

The cross-referencing of newly collected data with existing historical documents is a fundamental technique in the verification of data. Historical texts, maps, and records can provide a baseline against which new findings can be compared to. For example, architectural plans of historic buildings or old photographs can be used to validate current structural assessments or restoration projects. Comparing new data to these historical baselines allows researchers to identify and rectify inconsistencies and ensure a consistent and accurate representation of cultural heritage objects.

Another important technique is to verify data from multiple independent sources. This technique involves collecting similar data from different researchers, databases, or observations. The data is then compared for consistency. For example, the dimensions of a historic building recorded by one team can be verified against those recorded by another team at a different time. Another example is the comparison of multi-band 3D models derived from a variety of input data, such as RGB and IR images, UV-induced fluorescence images, etc., and/or acquired with different instruments (Grifoni et al. 2018). Discrepancies between these data sets can be used to identify potential errors or areas requiring further investigation. This multi-source verification process is particularly effective in building a comprehensive and reliable data set for heritage projects.

Data verification in cultural heritage documentation has been revolutionized by the use of advanced technologies such as laser scanning. Laser scanning provides high-resolution, three-dimensional representations of objects and sites (Ruther et al. 2012). These scans allow for precise measurements and detailed analysis. These scans can be compared to manual measurements or historical records for identification of inaccuracies. In addition, laser scanning can reveal details that are not visible to the naked eye, thus providing a more comprehensive understanding of the cultural heritage property.

The integration of these verification techniques; cross-referencing historical documents, corroborating multiple sources, and using laser scanning and GIS; creates a robust framework for ensuring data accuracy. This provides a multifaceted approach to verification, with each method complementing the others. This integrated approach not only improves the reliability of the data, but it also improves the overall understanding and preservation of the cultural heritage resource. As technology continues to advance, the tools and methods available for data verification will continue to evolve, allowing for even greater accuracy and reliability in the documentation of cultural heritage.

6.6 Data Documentation and Metadata

The proper documentation of data is essential for its effective use and re-use, especially in those areas where cultural heritage data are involved. Careful documentation of the context in which the data were collected is necessary when data are collected. This should include a record of the circumstances of data collection, such as the specific location, date, time, and conditions under which the data was collected. Such detailed documentation will ensure that future researchers have an understanding of the context and are able to make accurate interpretations. For example, knowing that a particular set of data was collected during a particular cultural heritage monument or at a particular time of year can have a significant impact on how that data is understood and used.

Another critical aspect that requires thorough documentation is the methodology used to collect data. Different methodologies can produce different results, and an understanding of the approach that was used is essential for anyone seeking to replicate the study or compare it with other research. Transparency and the ability to verify and validate the data can be achieved through detailed documentation of the methods used, including tools and techniques, sampling methods, and data processing steps.

This will also help to identify any potential biases or limitations in the data, which is crucial for the accurate analysis and interpretation of the data.

It is also important to clearly document the assumptions made in interpreting the data. These assumptions can have an impact on the conclusions that are drawn from the data and on the subsequent use of the data. By making these assumptions explicit, researchers provide a clear framework for understanding how the data have been interpreted and what potential biases or limitations may have been present. This level of transparency is essential for the credibility of the research and for enabling others to build on the work without misinterpretation.

Metadata plays an important role in the management and preservation of data, particularly in the area of cultural heritage. Metadata describes various aspects of the data, including its origin, purpose, temporal referencing, quality, format, and accessibility. This information is critical to the effective management of data, as it provides the necessary context and details that allow users to find, understand, and make appropriate use of the data. Properly structured metadata will ensure that the data can be discovered and accessed by others, thereby facilitating collaboration and further research.

Metadata not only facilitates data discovery and access. It also supports data preservation over time. Cultural heritage data often needs to be preserved for future generations. Metadata provides the information necessary to ensure that data remains usable and meaningful over long periods of time. By documenting the data's origin, context, and any changes made to it over time, metadata helps maintain the integrity and reliability of the data, making it a valuable resource for future research and preservation efforts.

In addition, the quality and reliability of data can be enhanced through well-documented metadata. Consistency and accuracy in how data is described and managed is ensured by metadata standards and best practices. This consistency allows for a standardized approach to data management, which is critical for integrating and comparing data from different sources. As a result, researchers will be able to more easily combine and compare data sets, resulting in more robust and comprehensive analyses. Proper data documentation and comprehensive metadata are essential components of effective data management to ensure that data can be used, re-used, and preserved for future generations.

6.7 Ethical Considerations

The acquisition and utilization of data in the realm of cultural heritage encompass a myriad of ethical considerations that are pivotal for upholding the integrity and credibility of the data. Chief among these concerns is the matter of privacy rights, particularly salient when handling sensitive information pertaining to individuals or communities linked to cultural heritage sites. Safeguarding personal data from unauthorized access and misuse stands as a paramount objective. Ethical protocols and legal frameworks such as the General Data Protection Regulation (GDPR) in Europe offer explicit guidelines for managing personal data and ensuring the preservation of privacy rights (EU 2016).

Furthermore, Intellectual Property Rights (IPRs) wield substantial influence over the ethical deliberations surrounding cultural heritage data. Intellectual property statutes safeguard creators' rights and guarantee due recognition and compensation for their contributions. This aspect assumes heightened significance when creative works like photographs, artwork, and literary pieces are involved in the data. Unauthorized use or dissemination of such creations can precipitate significant ethical and legal quandaries. Hence, securing appropriate permissions and licenses before incorporating intellectual property into cultural heritage initiatives is imperative.

A crucial ethical consideration that warrants attention is the potential exploitation of cultural heritage sites. Exploitation can manifest in various ways, such as the commercialization of these sites without providing equitable compensation to local communities or causing physical harm to the sites through irresponsible tourism and data collection techniques. It is imperative to adopt sustainable and respectful approaches to ensure the preservation of cultural heritage sites for future generations. This necessitates close collaboration with local authorities and stakeholders to devise strategies that safeguard these sites while allowing for their appreciation and scholarly exploration (Smith 2007).

Incorporating local communities and stakeholders into the data collection process is not only an ethical responsibility but also a pragmatic necessity. Local communities often hold invaluable knowledge and insights about cultural heritage sites, which can significantly enhance the quality and accuracy of the collected data. Moreover, involving them in the process fosters trust and ensures that data collection is carried out in a respectful and ethical manner. It also empowers these communities by granting them a sense of ownership and control over their cultural heritage (Smith 2007).

In order to uphold ethical standards in the collection and utilization of cultural heritage data, it is imperative to establish explicit guidelines and frameworks delineating the responsibilities and duties of all involved parties. This encompasses the establishment of review boards or ethics committees tasked with overseeing the data collection process, ensuring adherence to ethical standards, and addressing any ethical dilemmas that may arise. Moreover, continual education and training on ethical protocols for researchers and professionals in this field are essential to keep them abreast of the latest advancements and challenges.

Fundamentally, addressing the ethical considerations surrounding cultural heritage data collection and usage is pivotal for maintaining the credibility and integrity of such data. By acknowledging privacy rights, intellectual property, and the significance of cultural heritage sites while actively engaging local communities, researchers and practitioners can ensure that their endeavors contribute positively to preserving and appreciating cultural heritage. This ethical approach not only safeguards the heritage itself but also cultivates a more inclusive and respectful relationship between cultural heritage experts and the communities they serve (Smith 2007).

6.8 Challenges in Data Trustworthiness

The trustworthiness of cultural heritage data holds significant importance in ensuring precise historical interpretation, preservation endeavors, and educational utility. Nevertheless, various obstacles hinder the dependability of such data. Despite diligent efforts, challenges like technological constraints, human fallibility, and environmental influences substantially impact the quality of the data. Overcoming these hurdles necessitates continuous monitoring, data updates, and comprehensive training for personnel engaged in data collection and analysis.

Technological Constraints

A key impediment to maintaining trustworthy cultural heritage data is the limitation imposed by available technology. Sophisticated tools and methodologies are essential for accurately capturing, storing, and analyzing cultural heritage data. Unfortunately, not all cultural institutions have access to cutting-edge technology due to financial constraints or other resource limitations (Kalay et al. 2008). This can result in incomplete or inaccurate data, thereby compromising the overall quality and reliability of the information. For instance, imaging techniques with lower resolution may overlook crucial details, while inadequate storage solutions can lead to data deterioration over time.

The integration of advanced technologies such as big data analytics and digital documentation methods remains uneven across institutions. Many organizations are struggling with the high costs and technical expertise that are required to implement these systems in an effective manner. This technological divide means that some institutions are unable to benefit from the enhanced ability to manage and analyze data offered by modern technologies. As a result, these institutions cannot take full advantage of digital advances to protect and share their collections, and the preservation and interpretation of cultural heritage suffers (Poulopoulos and Wallace 2022).

The Impact of Human Error

Human error is a significant challenge that undermines the reliability of cultural heritage data. Errors can occur at various stages, from data collection to analysis and interpretation. Mistakes in data entry, mislabeling of artifacts, or improper use of technology can introduce inaccuracies that compromise the integrity of the data (Gillings et al. 2020). For example, during the digitization process, misidentification of objects or incorrect cataloging can lead to significant distortions in the recorded information, making it difficult to maintain accurate historical records. Moreover, human error is not limited to technical mistakes; subjective biases during data interpretation can also distort outcomes, leading to a flawed representation of cultural heritage.

To mitigate human error and increase the trustworthiness of cultural heritage data, it is imperative to implement comprehensive training programs and standardize procedures across institutions. Training programs can help ensure that individuals involved in data handling are well-versed in best practices, thereby reducing the likelihood of errors. In addition, standardizing data collection, processing, and storage methods can promote consistency and accuracy, making it easier to verify and validate data. By adopting these measures, cultural heritage institutions can improve the reliability of their data, ensuring that it accurately reflects the historical reality and supports sound research and preservation efforts.

Influence of Environmental Factors

Environmental factors play a crucial role in the deterioration and loss of cultural heritage data. Physical artifacts and digital storage media are particularly vulnerable to environmental conditions such as temperature, humidity, and light exposure (Hamilton and Crabbe 2009). For example, elevated humidity levels can promote mold growth on paper records, which can lead to irreversible damage and loss of historical information. Similarly, fluctuations in temperature can adversely affect electronic storage devices, potentially leading to data corruption or loss. Effective preservation strategies must therefore include rigorous environmental controls and regular monitoring to maintain optimal conditions and safeguard these invaluable resources.

Implementing robust preservation strategies to protect cultural heritage data from environmental hazards is essential. This involves the use of controlled environments with stable temperature and humidity levels, as well as protection from direct light exposure. Additionally, routine monitoring and maintenance of these conditions are necessary to prevent the gradual degradation of both physical and digital assets. The use of advanced technologies, such as climate control systems and environmental sensors, can help in maintaining the required conditions. Moreover, it is important to follow industry best practices for the storage and handling of cultural heritage materials to ensure their long-term preservation and accessibility for future generations (Guzman et al. 2018).

Continual Monitoring and Updating of Data

Implementing ongoing monitoring and updating protocols is essential to maintaining the trustworthiness of cultural heritage data. Data that is not regularly updated can quickly become obsolete or inaccurate. This is especially true as new information and technological advances emerge. In the context of digital records, technological advances can render previous formats and storage methods obsolete. This can lead to data loss or corruption if not managed properly. For example, the use of sensors and other advanced monitoring technologies has been effective in maintaining the integrity of cultural heritage sites by providing real-time data that can aid in the identification and mitigation of potential risks (Kasnesis et al. 2019).

The implementation of continuous monitoring requires not only the regular updating of data but also the adoption of new technologies that can improve the accuracy and relevance of data. For example, structural health monitoring (SHM) systems are essential for the assessment of the condition of historic structures and the planning of sustainable maintenance solutions (Rossi and Bournas 2023). Furthermore, integrating digital twins and graph-based data management allows for more detailed and dynamic tracking of cultural assets, ensuring that they remain well-documented and preserved for future generations (Falcone et al. 2021). Through the prioritization of these strategies, cultural heritage institutions can improve the reliability and usability of their data, contributing to better preservation and appreciation of cultural heritage.

Personnel Training and Education

The training and education of the personnel involved in the collection, analysis, and preservation of cultural heritage data is of critical importance in ensuring the accuracy and reliability of such data. Comprehensive training programs that cover the latest technologies, methodologies, and best practices can have a significant impact on the reduction of human error and the improvement of data quality. For example, to improve data integrity and consistency, professional training programs can teach staff how to effectively use advanced digital tools and software for documentation and preservation. Such training also ensures that staff are well versed in the theoretical aspects of cultural heritage preservation. This enables them to better understand the significance and context of the artifacts they work with (Achille and Fiorillo 2022).

In addition, in order to keep up with new challenges and technologies, it is essential to foster a culture of continuous learning and adaptation within cultural institutions. This includes not only initial training, but also ongoing opportunities for professional development that will allow staff to keep abreast of advances in the field. Regular workshops, seminars, and courses can help staff keep their skills and knowledge up to date and ensure that they are prepared to implement new techniques and respond to the evolving needs of the preservation community. By prioritizing education and continuous learning, institutions can maintain high standards of data management and preservation, ultimately helping to preserve cultural heritage for future generations (Jagielska-Burduk et al. 2021).

While there are significant challenges to maintaining the trustworthiness of cultural heritage data, a multifaceted approach that includes technological improvements, rigorous monitoring, and robust training can mitigate these issues. Key strategies for maintaining the integrity of cultural heritage information include addressing technological limitations, minimizing human error, protecting against environmental factors, and ensuring that data is continually updated. By prioritizing these areas, institutions can be more effective in the protection and preservation of cultural heritage for future generations.

References

Achille, C., and F. Fiorillo. 2022. Teaching and learning of cultural heritage: engaging education, professional training, and experimental activities. *Heritage* 5 (3). https://doi.org/10.3390/heritage5030134.

Alshawabkeh, Y., et al. 2020. Heritage documentation using laser scanner and photogrammetry. The case study of Qasr Al-Abidit, Jordan. *Digital Applications in Archaeology and Cultural Heritage* 16. https://doi.org/10.1016/j.daach.2019.e00133.

Commission, European, Content Directorate-General for Communications Networks, and Technology. 2022. Study on quality in 3D digitisation of tangible cultural heritage: mapping parameters, formats, standards, benchmarks, methodologies, and guidelines. VIGIE 2020/654—Final Study Report. https://doi.org/10.2759/471776.

EU. 2016. General Data Protection Regulation (GDPR). https://gdpr.eu/. Accessed 10 Jun 2024.

Falcone, M., et al. 2021. From architectural survey to continuous monitoring: graph-based data management for cultural heritage conservation with digital twins. *International Archives of the Photogrammetry, Remote Sensing and Spatial Information Sciences* XLIII-B4-2021: 47–53. https://doi.org/10.5194/isprs-archives-XLIII-B4-2021-47-2021.

Gillings, M., P. Haciguzeller, and G. Lock. 2020. *Archaeological spatial analysis–a methodological guide*, 512. Routledge. ISBN: 978-0815373230.

Grifoni, E., et al. 2018. Construction and comparison of 3D multi-source multiband models for cultural heritage applications. *Journal of Cultural Heritage* 34: 261–267. https://doi.org/10.1016/j.culher.2018.04.014.

Guzman, P., A. Roders, and B. Colenbrander. 2018. Impacts of common urban development factors on cultural conservation in world heritage cities: an indicators-based analysis. *Sustainability* 10 (3). https://doi.org/10.3390/su10030853.

Hamilton, R., and H. Crabbe. 2009. Environment, pollution and effects. In *The effects of air pollution on cultural heritage*, ed. R. Hamilton, et al., 1–27. Boston, MA: Springer US. ISBN: 978-0-387-84893-8. https://doi.org/10.1007/978-0-387-84893-8_1.

Jagielska-Burduk, A., M. Pszczynski, and P. Stec. 2021. Cultural heritage education in UNESCO cultural conventions. *Sustainability* 13 (6). https://doi.org/10.3390/su13063548.

Jokilehto, J. 2006. Considerations on authenticity and integrity in world heritage context. City and Time 2 (1).

Kalay, Y., T. Kvan, and J. Affleck. 2008. *New heritage: new media and cultural heritage*, 1st ed., 340. Routledge. ISBN: 978-0415773560.

Kasnesis, P., et al. 2019. Acoustic sensor data flow for cultural heritage monitoring and safeguarding. *Sensors* 19 (7). https://doi.org/10.3390/s19071629.

Letellier, R. 2007. *Recording, documentation, and information management for the conservation of heritage places*, 151. Guiding Principles. The Getty Conservation Institute. ISBN: 978-0-89236-925-6.

Moody, B., et al. 2021. Digital chronological data reuse in archaeology: Three case studies with varying purposes and perspectives. *Journal of Archaeological Science: Reports* 40: 103188. https://doi.org/10.1016/j.jasrep.2021.103188.

Poulopoulos, V., and M. Wallace. 2022. Digital technologies and the role of data in cultural heritage: The past, the present, and the future. *Big Data and Cognitive Computer* 6 (3). https://doi.org/10.3390/bdcc6030073.

Prechtel, N., and S. Munster. 2016. Cultural heritage in a spatial context–towards an integrative, interoperable, and participatory data and information management. In *3D research challenges in cultural heritage II—how to manage data and knowledge related to interpretative digital 3D reconstructions of cultural heritage*, ed. S. Munster et al., 272–278. https://doi.org/10.1007/978-3-319-47647-6_13.

Rossi, M., and D. Bournas. 2023. Structural health monitoring and management of cultural heritage structures: a state-of-the-art review. *Applied Sciences* 13 (11). https://doi.org/10.3390/app13116450.

Ruther, H., et al. 2012. Laser scanning in heritage documentation: the scanning pipeline and its challenges. *Photogrammetric Engineering and Remote Sensing* 4: 309–316. https://doi.org/10.14358/PERS.78.4.309.

Smith, L. 2007. *Uses of heritage*, 368. Routledge. ISBN: 978-0415318310.

Vecco, M. 2010. A definition of cultural heritage: From the tangible to the intangible. *Journal of Cultural Heritage* 1 (3): 321–324. https://doi.org/10.1016/j.culher.2010.01.006.

Chapter 7
Legal Aspects

Abstract This chapter highlights the importance of cultural heritage recording and documentation, and the critical role of legal frameworks in the protection of cultural artifacts, sites, and practices. The integrity and ownership of cultural expressions are ensured by intellectual property laws, repatriation laws, and the protection of intangible heritage. National and international laws, such as the UNESCO Conventions, guide global cooperation. This chapter examines intellectual property rights. It focuses on copyright, moral rights, and related rights to promote responsible documentation. Data protection, informed consent, and anonymity are critical to respect for the privacy of individuals. Ownership and custodianship involve complex legal principles. They aim to prevent illicit trafficking and ensure legitimate ownership. A comparative analysis of national legal frameworks reveals challenges and opportunities. It advocates for harmonization, international cooperation, and technological advances while balancing ethical considerations with legal compliance.

7.1 Introduction

The recording and documentation of cultural heritage is a critical task in preserving the history and identity of communities and nations. The legal aspects that surround these processes play a vital role in ensuring that cultural heritage is protected, respected, and accurately represented. Legal frameworks provide the necessary guidelines and protections to prevent cultural monuments, sites, artifacts and practices from becoming misappropriated, misrepresented, or destroyed. As such, an understanding of the legal aspects of the documentation of cultural heritage is essential for professionals who work in the field (Blake 2015).

Legislation and legal issues intersect with the processes of recording and documentation of cultural heritage in a number of ways. Intellectual property laws protect the rights of those creating and preserving cultural expressions. These laws ensure that those who have produced or maintained cultural heritage have control over how it is used and how it is shared. In addition, cultural property repatriation laws address the return of artifacts to their places of origin, the redress of historical injustices, and the restoration of cultural integrity (Francioni and Gordley 2013). Legal protection

E. Stylianidis, *Exploring the Ethical Dimension in Recording and Documenting Cultural Heritage*, https://doi.org/10.1007/978-3-031-80034-4_7

also extends to preserving intangible cultural heritage, such as oral histories and traditional knowledge, to protect it from exploitation and loss (UNESCO 2003).

National laws and international agreements have a significant impact on cultural heritage recording and documentation. National legislation determines how cultural heritage is to be managed within a country, defining what constitutes heritage and the responsibilities of individuals and institutions in its conservation (Blake 2015). International agreements provide a framework for global cooperation and standard-setting, such as the UNESCO Convention for the Safeguarding of the Intangible Cultural Heritage (UNESCO 2003) and the Hague Convention for the Protection of Cultural Property in the Event of Armed Conflict (UNESCO 1954). These conventions help to harmonize efforts across borders, to ensure that cultural heritage is protected on a global scale, and to foster a sense of shared responsibility (O'Keefe 2006).

The legal aspects of recording and documenting cultural heritage are essential to preserving and protecting cultural heritage. Through the provision of clear guidelines and protections, legal frameworks help to ensure that cultural heritage is accurately recorded and respectfully managed. Understanding these legal dimensions is critical for anyone involved in cultural heritage work, as it helps to navigate the complexities of owning, accessing, and preserving in a manner that honors the cultural significance of the heritage being documented.

7.2 Intellectual Property Rights

7.2.1 Copyright

Copyright is an essential element of IPR, which protects the original work of writers, artists, and organizations and ensures control over the use and distribution of the work. In the context of cultural heritage, copyright covers a broad range of works, including literature, music, art, architecture, and even certain documented and conserved performances. The primary goal of copyright law is the encouragement of creativity by providing creators with exclusive rights in their works for a period of time, typically the life of the author plus 50–70 years posthumously (WIPO 2016). This protection ensures that cultural heritage is documented, preserved, and shared responsibly while allowing creators to benefit financially and receive recognition for their work.

Copyright law grants several rights to authors, creators, and institutions involved in the creation and preservation of cultural heritage. These rights include the right to reproduce the work, the right to distribute copies, the right to perform or display the work publicly, and the right to prepare derivative works based on the original. For cultural heritage documentation, these rights are critical because they ensure that the creators or custodians of cultural heritage artifacts can control the use and distribution of these items and prevent unauthorized exploitation. After the copyright expires, the work is usually in the public domain. This transition has important implications for the documentation of cultural heritage, as it requires a balance between the protection of the rights of creators and the broader public interest in access to and preservation of cultural heritage (EC 2019).

7.2.2 Moral Rights

Moral rights are a subset of copyright. They protect the personal and reputational interests of creators. These rights are crucial in the field of cultural heritage because they ensure that creators are properly credited for their work and that their works are not subjected to disparaging treatment that may harm their reputations. Moral rights typically include the right to attribution, which ensures that the creator is recognized as the author of the work, and the right to integrity, which allows creators to object to any distortion, mutilation, or other alteration of their work that might harm their honor or reputation (WIPO 2016).

Several case studies illustrate the importance of moral rights in the context of cultural heritage. For example, the controversy surrounding the unauthorized alteration of historical monuments or works of art often hinges on the violation of the right to integrity of the artist. Similarly, in the case of the misappropriation or misrepresentation of Indigenous cultural expressions, the right to attribution is a critical tool to ensure that the creators and their communities receive proper recognition and respect (Brown 2004). These rights help maintain the authenticity and integrity of cultural heritage. They also promote respect for the creators and their cultural context.

7.2.3 Related Rights

Related rights, also known as neighboring rights, extend copyright-like protection to other contributors who are involved in the creation and dissemination of cultural expressions. These rights are particularly relevant to performers and broadcasters, who play an important role in preserving and promoting intangible cultural heritage (WIPO 2016). The rights of performers protect the individuals who perform literary and artistic works and provide them with control over the recording, reproduction, and broadcasting of their performances. This is critical to the protection of the interests of performers who contribute to cultural heritage through live performances so that they can be compensated and recognized for their contributions.

Similarly, for the preservation and dissemination of cultural heritage, the rights of phonogram producers and broadcasters are essential. Producers of phonograms have rights that protect their investment in the recording and distribution of sound works. These rights ensure that they can control and profit from the use of their recordings. Broadcasters, on the other hand, have rights that protect their broadcasts from unauthorized use. These rights enable them to control the distribution and reproduction of their programs (WIPO 2016). These related rights ensure that all contributors to cultural heritage, from performers to producers, are adequately protected and incentivized to continue their efforts in the preservation and promotion of cultural heritage.

7.3 Data Protection and Privacy

7.3.1 Personal Data in Cultural Heritage Documentation

The handling of personal data is subject to strict legal requirements designed to protect the privacy of individuals in the field of cultural heritage documentation. GDPR in the European Union is one such regulation, mandating the lawful processing of personal data. According to the GDPR, personal data must be processed in a lawful, fair, and transparent manner. Organizations involved in cultural heritage documentation must ensure that their data collection practices adhere to these principles, incorporating safeguards such as data minimization and purpose limitation.

Privacy concerns in cultural heritage documentation go beyond compliance with legal requirements. Ethical considerations play a critical role in ensuring that the process of documenting respects the dignity and rights of individuals. This includes being mindful of the context in which personal information is being collected and the potential impact of the disclosure of that information. Ethical guidelines often emphasize the need for trust and respect in the documentation process and advocate for transparency with participants about how their data will be used, stored, and shared.

7.3.2 Anonymity and Consent

Informed consent is a keystone of ethical cultural heritage recording. Informed consent means that participants understand the nature of the documentation process, the type of data collected, and how it will be used. This process needs to be clear and straightforward. Technical jargon that may confuse participants should be avoided. Legal frameworks, such as the GDPR in Europe, provide strong support for informed consent and ensure that participants' rights to privacy and data protection are upheld.

The importance of anonymity in the documentation of cultural heritage cannot be overstated. Anonymization of data helps to protect the identities of individuals, thereby reducing the risk of privacy violations. This is often achieved through techniques such as pseudonymization and data masking. In addition, participants should be informed of their right to withdraw consent at any time, which helps to empower them to control their personal data. Jurisdictions around the world recognize the importance of consent and anonymity. They have incorporated these principles into their legal frameworks to protect personal information in cultural heritage documentation.

7.4 Ownership and Custodianship

7.4.1 Ownership

The ownership of cultural heritage is governed by a complex framework of legal principles that vary from one jurisdiction to another, but often share common elements. International conventions, such as the UNESCO Convention of 1970, play an important role in shaping national laws relating to owning and protecting cultural property. This Convention seeks, among other things, to prevent the illicit trafficking of cultural property and to ensure the return of cultural property that has been stolen or illegally exported to its rightful owners (UNESCO 1970). The Convention on Stolen or Illegally Exported Cultural Objects (UNIDROIT 1995), supplements that of 1970 in terms of private law. The concept of ownership goes beyond the mere possession of cultural property; it encompasses the rights to control, use, and enjoy the benefits of cultural property (O'Keefe 1997).

In contexts where cultural artifacts have been removed from their countries of origin, sometimes during colonial times, and are now held in foreign museums or private collections, disputes over the ownership of cultural heritage often arise. Diplomatic negotiations, legal action, and international arbitration can be used to resolve these disputes. High-profile cases such as the dispute between Greece and the United Kingdom over the Parthenon Sculptures, for example, highlight the ongoing challenges of reconciling historical grievances with the legal frameworks of the present day (Greenfield 2007). Courts and international bodies are often called upon to balance the interests of the original owners, the current possessors, and the public's access to cultural heritage.

7.4.2 Custodianship and Stewardship

Custodianship and stewardship of cultural heritage refer to the roles and responsibilities of entities or individuals entrusted with caring for, preserving, and managing cultural property. The custodians of cultural heritage may include museums, government agencies, private collectors, and indigenous communities. Their primary responsibility is the preservation of cultural heritage for future generations, which includes the maintenance of the integrity, authenticity, and accessibility of cultural objects and sites (UNESCO 2003).

Legal agreements and frameworks support the role of custodians through a variety of means, including national laws, international treaties, and ethical guidelines. For example, the 2003 UNESCO Convention for the Safeguarding of the Intangible Cultural Heritage emphasizes the role of communities in the protection of their cultural practices and traditions, in recognition of the fact that effective custodianship often requires local involvement and expertise (UNESCO 2003).

Custodianship also involves legal and ethical considerations related to exhibiting, preserving, and returning cultural heritage. Institutions that hold cultural artifacts are increasingly adopting policies that reflect a commitment to transparency, cultural sensitivity, and proactively returning items to rightful owners when appropriate. This shift toward ethical stewardship recognizes that the protection of cultural heritage is not only a legal obligation but also a moral one. It fosters a greater sense of global responsibility and cultural understanding (O'Keefe and Prott 2011).

7.5 National Legal Frameworks

7.5.1 Comparative Analysis

Reflecting the unique historical, cultural and legal context of each country, national laws on cultural heritage documentation show considerable diversity. For example, Italy places a strong emphasis on the preservation and documentation of cultural heritage as a fundamental duty of the state. This comprehensive framework mandates detailed inventories and regular updates to ensure the preservation of cultural assets. In contrast, the United States operates under a more decentralized system, where heritage laws provide a framework that allows individual states to adopt specific documentation standards. This approach results in a diverse landscape of cultural heritage documentation practices across the country, reflecting local priorities and resources.

In addition, Japan integrates heritage documentation into broader cultural policies that emphasize community engagement and the use of advanced technologies. Japan requires meticulous documentation and engages local communities in the preservation process, fostering a sense of shared responsibility. Similarly, Australia emphasizes the role of both the federal and state governments in the documentation and preservation of cultural heritage. This two-tiered approach ensures that documentation efforts are comprehensive and inclusive, taking into account both the national heritage and the heritage of local communities.

Launched in 2005 and funded by the U.S. Department of State, the UNESCO Database of National Cultural Heritage Laws provides free and easy access to current cultural heritage laws and regulations from 188 Member States, covering 2756 laws. This tool is invaluable for stakeholders such as governments, customs officials, art dealers, and lawyers. It provides a comprehensive source of information on national cultural heritage laws, export/import certificates, translations of laws, and contact information for national authorities. In addition, a quarterly newsletter informs users of the latest changes to the database and provides information on upcoming workshops and meetings at which the database will be promoted (UNESCO 2023b).

7.5.2 Challenges and Opportunities

Countries have a number of common challenges with regard to legislation on cultural heritage documentation. A key issue is the tension between development and conservation, where economic interests are often at odds with the need for documentation and protection of cultural sites (Kostakis and Lolos 2024). Furthermore, the lack of standardized documentation practices can lead to inconsistencies and gaps in documentation, complicating international cultural heritage protection and management efforts. These challenges are further exacerbated, particularly in developing countries, by financial constraints and limited access to advanced documentation technologies. Another important aspect is crowdfunding for cultural heritage projects (Jelincic and Sveb 2021).

There are, however, significant opportunities for harmonization and international collaboration. International frameworks, such as the UNESCO World Heritage Convention, provide a basis for the standardization of documentation practices across borders and promote more cohesive global efforts for the protection of cultural heritage. In addition, technological advances such as digital archiving and 3D scanning provide innovative solutions to documentation challenges, allowing for more comprehensive and accessible records (Kantaros et al. 2023). Collaborative projects and knowledge exchange between countries can also enhance national capacity and foster a shared commitment to the preservation of cultural heritage for future generations.

7.6 Ethical Considerations and Legal Compliance

7.6.1 Balancing Legal and Ethical Concerns

Navigating the intersection of legal compliance and ethical practices in documentation presents both challenges and opportunities. Legal requirements are often clear, mandating specific actions or prohibitions to ensure compliance with established laws and regulations. However, ethical considerations can be more nuanced and subjective. They are guided by principles of fairness, integrity, and respect for all stakeholders. Legal and ethical obligations often align, providing a straightforward path to compliant and morally sound documentation practices.

Conflicts can arise, however, when legal mandates do not take ethical considerations fully into account. This tension is illustrated by a case study in the technology industry. Certain data collection practices may be legally permissible under current laws. However, they raise significant ethical concerns about user privacy and consent. For example, while a social media site may legally collect user data to improve its offerings, the ethical question of transparency in the collection and use of that data remains controversial. Scenarios such as these underscore the importance for organizations to take a proactive stance that goes beyond mere compliance with the law and addresses the broader ethical implications.

7.6.2 Best Practices

Several best practices can help ensure both compliance and ethical integrity when navigating the complex landscape of legal and ethical documentation. First, organizations should implement comprehensive training programs. These programs should stress the importance of understanding both legal requirements and ethical standards. This dual focus will ensure that employees are equipped not only to comply with the law, but also to be proactive in recognizing and dealing with ethical dilemmas.

Second, the establishment of clear and transparent policies is of critical importance. These policies should outline the organization's commitment to ethical practices. They should also provide guidance on how to handle situations where legal and ethical considerations may conflict. For example, the organizations should adopt a privacy policy that not only complies with legal standards but also clearly communicates to users how their information will be used and how it will be protected. Regular audits and reviews of these policies may ensure that they remain relevant and effective as the legal and ethical environment evolves.

Finally, it is essential to foster a culture of open dialogue and ethical reflection within the organization. More informed and morally sound decisions can be made when employees are encouraged to voice concerns and participate in discussions about ethical issues. This inclusive approach contributes to the development of a robust ethical framework that supports compliance with the law and at the same time respects the broader values and expectations of society.

References

Blake, J. 2015. *International cultural heritage law*, 384. Oxford University Press. ISBN: 9780198723516.

Brown, M. 2004. *Who owns native culture*?, 336. Harvard University Press. ISBN: 978-0674016330.

EC. 2019. Directive (EU) 2019/790 of the European Parliament and of the Council of 17 April 2019 on copyright and related rights in the Digital Single Market and amending Directives 96/9/EC and 2001/29/EC. https://eur-lex.europa.eu/legal-content/EN/TXT/PDF/?uri=CELEX:32019L0790. Accessed 25 Jun 2024.

Francioni, F., and J. Gordley. 2013. *Enforcing international cultural heritage law*, 280. Oxford University Press. ISBN: 9780199680245.

Greenfield, J. 2007. *The return of cultural treasures*, 3rd ed., 522. Cambridge University Press. ISBN: 978-0521802161.

Jelincic, D., and M. Sveb. 2021. Financial sustainability of cultural heritage: A review of crowdfunding in Europe. *Journal of Risk and Financial Management* 14 (3). https://doi.org/10.3390/jrfm14030101.

Kantaros, A., T. Ganetsos, and F. Petrescu. 2023. Three-dimensional printing and 3D scanning: Emerging technologies exhibiting high potential in the field of cultural heritage. *Applied Sciences* 13 (8). https://doi.org/10.3390/app13084777.

Kostakis, I., and S. Lolos. 2024. Uncovering the impact of cultural heritage on economic growth: empirical evidence from Greek regions, 2000–2019. *The Annals of Regional Science*. https://doi.org/10.1007/s00168-024-01280-3.

O'Keefe, P., and L. Prott. 2011. *Cultural heritage conventions and other instruments: a compendium with commentaries*, 343. ISBN: 978-1-903987-12-4.

O'Keefe, P. 1997. *Trade in antiquities: Reducing destruction and theft*, 134. ISBN: 978-92-3-103406-0.

O'Keefe, P. 2006. *The protection of cultural property in armed conflict*, 432. Cambridge University Press. ISBN: 9780521867979.

UNESCO. 1954. 1954 Convention for the Protection of Cultural Property in the Event of Armed Conflict. https://en.unesco.org/protecting-heritage/convention-and-protocols/1954-convention. Accessed 05 Jan 2024.

UNESCO. 1970. Convention on the Means of Prohibiting and Preventing the Illicit Import, Export and Transfer of Ownership of Cultural Property. https://unesdoc.unesco.org/ark:/48223/pf0000133378. Accessed 29 Jun 2024.

UNESCO. 2003. Text of the Convention for the Safeguarding of the Intangible Cultural Heritage. https://ich.unesco.org/en/convention. Accessed 03 May 2024.

UNESCO. 2023b. UNESCO Database of National Cultural Heritage Laws. https://www.unesco.org/en/articles/unesco-database-national-cultural-heritage-laws-updated. Accessed 30 Jun 2024.

UNIDROIT. 1995. The Convention on Stolen or Illegally Exported Cultural Objects. https://www.unidroit.org/instruments/cultural-property/1995-convention/. Accessed 29 Jun 2024.

WIPO. 2016. Understanding Copyright and Related Rights. https://www.wipo.int/edocs/pubdocs/en/wipo_pub_909_2016.pdf. Accessed 25 Jun 2024.

The manufacturer's authorised representative in the EU is Springer Nature Customer Service Centre GmbH, Europaplatz 3, 69115 Heidelberg, Germany. If you have any concerns regarding our products, please contact ProductSafety@springernature.com

Printed and bound by CPI Group (UK) Ltd, Croydon, CR0 4YY
30/09/2026
02225607-0008